E-MRS 2012 Spring Meeting, Symposium M: More than Moore: Novel Materials Approaches for Functionalized Silicon Based Microelectronics

IOP Conference Series: Materials Science and Engineering
Volume 41

Strasbourg, France
14 - 18 May 2012

ISBN: 978-1-63439-852-7
ISSN: 1757-8981

Printed from e-media with permission by:

Curran Associates, Inc.
57 Morehouse Lane
Red Hook, NY 12571

proceedings.com

Some format issues inherent in the e-media version may also appear in this print version.

Printed by Curran Associates, Inc. (2015)

Institute of Physics
Dirac House, Temple Back
Bristol BS1 6BE UK

Phone: 44 1 17 929 7481
Fax: 44 1 17 920 0979

techtracking@iop.org

Additional copies of this publication are available from:

Curran Associates, Inc.
57 Morehouse Lane
Red Hook, NY 12571 USA
Phone: 845-758-0400
Fax: 845-758-2634
Email: curran@proceedings.com
Web: www.proceedings.com

Table of contents

Volume 41

E-MRS 2012 Spring Meeting, Symposium M: More than Moore: Novel materials approaches for functionalized Silicon based Microelectronics
14–18 May 2012, Strasbourg, France

Accepted papers received: 12 September 2012
Published online: 6 December 2012

Papers

PREFACE

More than Moore explores a new area of Silicon based microelectronics, which reaches beyond the boundaries of conventional semiconductor applications. Creating new functionality to semiconductor circuits, More than Moore focuses on motivating new technological possibilities.

In the past decades, the main stream of microelectronics progresses was mainly powered by Moore's law, with two focused development arenas, namely, IC miniaturization down to nano scale, and SoC based system integration. While the microelectronics community continues to invent new solutions around the world to keep Moore's law alive, there is increasing momentum for the development of 'More than Moore' technologies which are based on silicon technologies but do not simply scale with Moore's law.

Typical examples are RF, Power/HV, Passives, Sensor/Actuator/MEMS or Bio-chips. The More than Moore strategy is driven by the increasing social needs for high level heterogeneous system integration including non-digital functions, the necessity to speed up innovative product creation and to broaden the product portfolio of wafer fabs, and the limiting cost and time factors of advanced SoC development. It is believed that More than Moore will add value to society on top of and beyond advanced CMOS with fast increasing marketing potentials.

Important key challenges for the realization of the 'More than Moore' strategy are:

- perspective materials for future THz devices
- materials systems for embedded sensors and actuators
- perspective materials for epitaxial approaches
- material systems for embedded innovative memory technologies
- development of new materials with customized characteristics

The Hot topics covered by the symposium M (More than Moore: Novel materials approaches for functionalized Silicon based Microelectronics) at E-MRS 2012 Spring Meeting, 14–18 May 2012 have been:

- development of functional ceramics thin films
- New dielectric materials for advanced microelectronics
- bio- and CMOS compatible material systems
- piezoelectric films and nanostructures
- Atomic Layer Deposition (ALD) of oxides and nitrides
- characterization and metrology of very thin oxide layers

We would like to take this opportunity to thank the Scientific Committee and Local Committee for bringing together a coherent and high quality Symposium at E-MRS 2012 Spring Meeting.

Christian Wenger, Jean Fompeyrine, Christophe Vallée and Jean-Pierre Locquet
Organizing Committee of Symposium M
September 2012

IOP Conf. Series: Materials Science and Engineering **41** (2012) 012001 doi:10.1088/1757-899X/41/1/012001

A new concept for spatially divided Deep Reactive Ion Etching with ALD-based passivation

F. Roozeboom[1,2*], B. Kniknie[1], A.M. Lankhorst[1], G. Winands[1], R. Knaapen[1], M. Smets[1], P. Poodt[1], G. Dingemans[2], W. Keuning[2] and W.M.M. Kessels[2]
[1] TNO, PO Box 6235, 5600 HE Eindhoven, The Netherlands
[2] Eindhoven University of Technology, PO Box 513, 5600 MB Eindhoven, The Netherlands

E-mail: f.roozeboom@tue.nl ; fred.roozeboom@tno.nl

Abstract. Conventional Deep Reactive Ion Etching (DRIE) is a plasma etch process with alternating half-cycles of 1) Si-etching with SF_6 to form gaseous SiF_x etch products, and 2) passivation with C_4F_8 that polymerizes as a protecting fluorocarbon deposit on the sidewalls and bottom of the etched features. In this work we report on a novel alternative and disruptive technology concept of Spatially-divided Deep Reactive Ion Etching, S-DRIE, where the process is converted from the time-divided into the *spatially divided* regime. The spatial division can be accomplished by inert gas bearing 'curtains' of heights down to ~20 μm. These curtains confine the reactive gases to individual (often linear) injection slots constructed in a gas injector head. By horizontally moving the substrate back and forth under the head one can realize the alternate exposures to the overall cycle.
A second improvement in the spatially divided approach is the replacement of the CVD-based C_4F_8 passivation steps by ALD-based oxide (e.g. SiO_2) deposition cycles. The method can have industrial potential in cost-effective creation of advanced 3D interconnects (TSVs), MEMS manufacturing and advanced patterning, e.g., in nanoscale transistor line edge roughness using Atomic Layer Etching.

Introduction

3D through-silicon vias (TSVs) date back to two patents in the 1960s [1, 2], *cf.* Fig. 1. Yet, it is only now with the continuous on-chip scaling reaching the point where Moore's Law (essentially an economic law) approaches its limits that TSV technology receives increasing interest for 3D integration [3]. Other drivers next to cost reduction are the reduced form factor and the increased performance of TSV-connected stacked-die devices, such as reduced RC delay and low power consumption. TSV technology is also accelerating the rapidly growing market of microelectromechanical systems (MEMS) by enabling the interconnection of multifunctional chips stacked in a heterogeneously 3D-integrated System-in-Package (*i.e.* the so-called '*More than Moore*' domain). Today, the industrial technology of choice for etching both TSV and MEMS structures in silicon is Deep Reactive Ion Etching (DRIE).

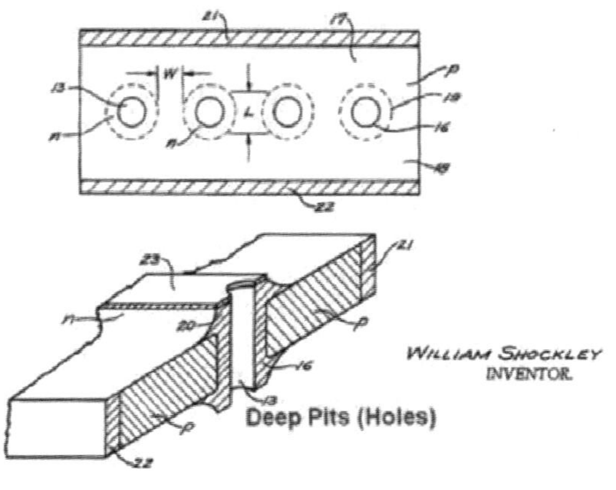

Figure 1. TSV structures proposed in Shockley's patent [1].

Spatially divided Deep Reactive Ion Etching: a new concept

The conventional technology of choice for silicon DRIE etching is the room temperature Bosch process [4, 5] illustrated in Fig. 2a. This process consists of two alternating half-cycles: 1) etching with SF_6 plasma, and 2) passivation of the sidewalls and bottom of the etched features with a protecting -$(C_2F_4)_n$- fluorocarbon, PTFE-like) polymer liner deposited from C_4F_8 plasma.

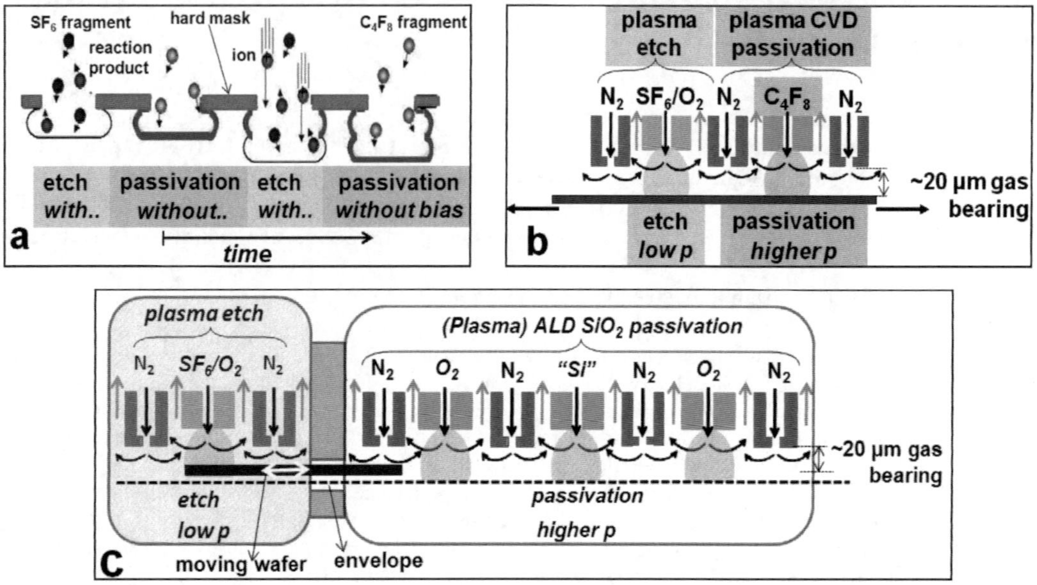

Figure 2. a) Conventional Bosch etch process scheme with temporal switching of consecutive etch and passivation half-cycles. The horizontal bar in grey represents a pre-patterned hard mask; (b) alternative spatial process modes with C_4F_8 passivation; (c) alternative with spatial ALD SiO_2 passivation of a wafer which moves horizontally back and forth under spatially divided reaction zones. Blue arrows pointing upwards indicate exhaust lines. Notice the difference in height of the gas bearing compartments (down to ~20 μm) and the plasma compartments (order ~mm); not to scale.

The first half-cycle is an ion-assisted isotropic etch step with SF_6 plasma. It would proceed - if non-interrupted - mainly by the non-directional F-containing radicals to form volatile SiF_x products that are pumped off. In order to minimize the lateral etching component the etch steps are quickly interrupted by C_4F_8 passivation steps. During each etch step a bias voltage is applied to the substrate holder. This causes a directional physical ion bombardment from the plasma onto the substrate which sputters the polymer off the feature's bottom part, thus leaving the sidewall passivation intact, and enabling the anisotropic etching.

The etch and passivation cycle times are each typically 1-10 s with 0.1-1 μm etched per cycle. The process enables plasma etching of deep vertical microstructures (aspect ratios AR ≥ 20:1) in silicon with etch rates of typically 3-5 μm/min, and selectivities up to ~200:1 against a hard oxide mask (usually SiO_2) [6].

An accelerated etch alternative is to convert the above process from its temporal (i.e. *time-separated*) into the *spatially separated* regime [7]. The spatial separation can be accomplished by inert gas (e.g. N_2) bearing 'curtains' of heights down to ~20 μm, or even smaller (Fig. 2b). These curtains confine the reactive gases to individual (often linear) injection zones constructed in a gas injector head. By horizontally moving the substrate back and forth under the multiple injector head one can create the alternate exposures needed to complete the overall cycle. The optimum pressure in each injection slot is obtained by balancing the various gas flows which are injected into and exhausted from the slots, and by a proper design of the distance between the various slots and the gas bearing gap height (a smaller gap causes a larger pressure field gradient between the various channels).

The passivation step in spatial DRIE: ALD-based, low-pressure or atmospheric

The selected mask material generally affects etch rate, undercutting, and surface quality of etched features [8]. Oxidic ALD-deposited hard masks like Al_2O_3 are reported to have lower pinhole density and thus superior etch selectivity than conventionally deposited etch hard masks [9, 10]. Thus a further improvement in the spatial approach can be expected from the replacement of the CVD-based C_4F_8 passivation steps by ALD-based deposition cycles of SiO_2, or other oxides (e.g. Al_2O_3). Unlike the C_4F_8 case the ALD-based passivation layer is self-limiting and chemisorptive of nature, and less complex in its layer thickness control. This will lead to improved control of the anisotropy and sidewall smoothness in the total DRIE process.

The idea of using temporal ALD passivation in DRIE of high aspect ratio nanosize features was conceived recently [11], yet without any experimental data given. Experimental evidence was reported by Dingemans *et al.* [12]. They published a time-efficient plasma-assisted process for low-temperature (50-400 °C) temporal ALD of SiO_2 using $H_2Si[N(C_2H_5)_2]_2$ precursor known as SAM.24, and O_2-plasma. Precursor dosing times as short as ~50 ms were sufficient to obtain a high conformality (95 ± 5 %) over high aspect ratio (30:1) trenches, as illustrated in Fig. 3a. This indicates that the recombination of O-radicals in such trenches plays no dominant role as was also discussed recently by Knoops *et al.* [13].

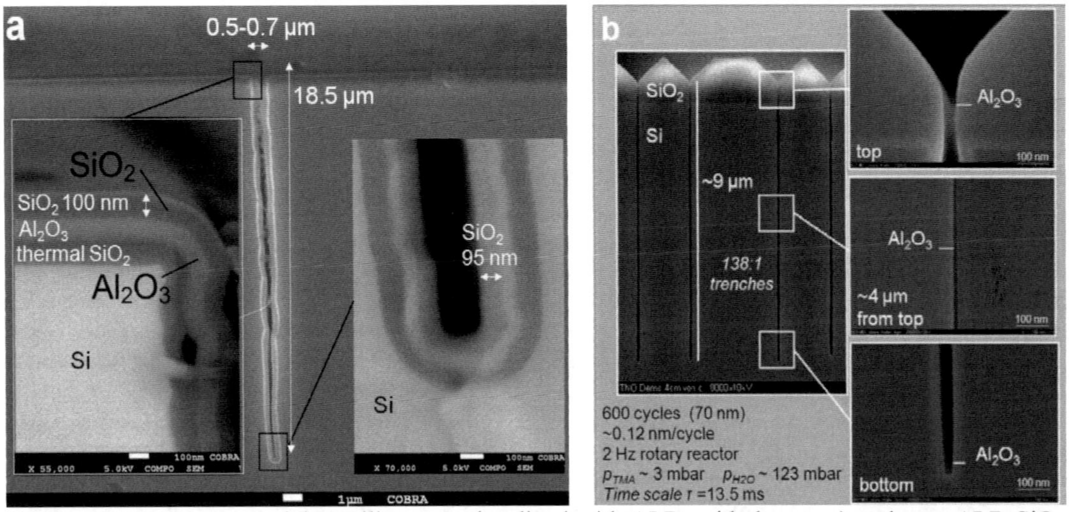

Figure 3. SEM images of deep silicon trenches lined with ALD oxide layers: a) a plasma ALD SiO_2 layer deposited at low-pressure in trenches with aspect ratio ~30 during 830 cycles on top of an ALD Al_2O_3 /thermal SiO_2 layer stack inside. Note, that the wavy appearance of the full trench is due a sample cleaving artefact; (after ref. [12] © Electrochemical Society 2012); b) an ALD Al_2O_3 layer deposited at 1 atm. in 138:1 aspect ratio trenches during 600 cycles in a rotary ALD reactor (trenched wafers kindly provided by Fraunhofer CNT/Namlab, Dresden). Note: in an actual DRIE application the passivation would require only a few ALD cycles (i.e. monolayers of SiO_2 or Al_2O_3).

We found that a non-plasma *atmospheric* spatial ALD alternative for oxidic passivation is also possible. Figure 3b shows such an Al_2O_3 layer deposited from trimethyl aluminum and water vapor during 600 repeated cycles of 13.5 ms each in the rotary atmospheric ALD reactor described earlier [14]. The layer has good step conformality (≥ 80 %) in trenches with ultrahigh aspect ratios exceeding 130:1. Atmospheric ALD of Al_2O_3 has already successfully been commercialized for the solar cell industry in equipment that deposits films almost two orders of magnitude faster than in conventional (temporal) ALD [15]. This opens up the way to the development of ultrafast atmospheric passivation in DRIE, which would not only simplify and accelerate etching, but also reduce costs (sub-second passivation cycles with, for example, Al_2O_3 depositing at 1.2 nm/s).

Spatial DRIE Reactor Design

The basics of a full spatial DRIE process scheme are illustrated in Fig. 2c. Figure 4 gives an impression of the basics in spatial reactor gas inlet design: a wafer is moving under a (plasma) injector head with inlets for etch gas (SF_6/O_2), bearing gas (N_2) and passivation gas (conventional C_4F_8 or ALD oxide). The pressures p_e, p_p and p_{pu} assumed for the etch, passivation and purge zones, respectively, and the corresponding flow rates Φ_e and Φ_p, lengths L_e and L_p and heights H_e, H_p and H_g of the injection zones are listed in Fig. 4. H is a convenient design parameter to obtain the desired pressures. The pressure drop over each channel is proportional to the cube of its compartment height ($\Delta p \sim H^3$), and linear in L and Φ. Depending on the pressures needed for the spatial DRIE process one can calculate the different dimensions of the bearings. A typical example for low-pressure DRIE is shown in Figure 4, indicating these dimensions to be in the mm to sub-mm range. Note, that the pressure for passivation, p_p, is taken to be one order of magnitude higher than the pressure p_e for etching.

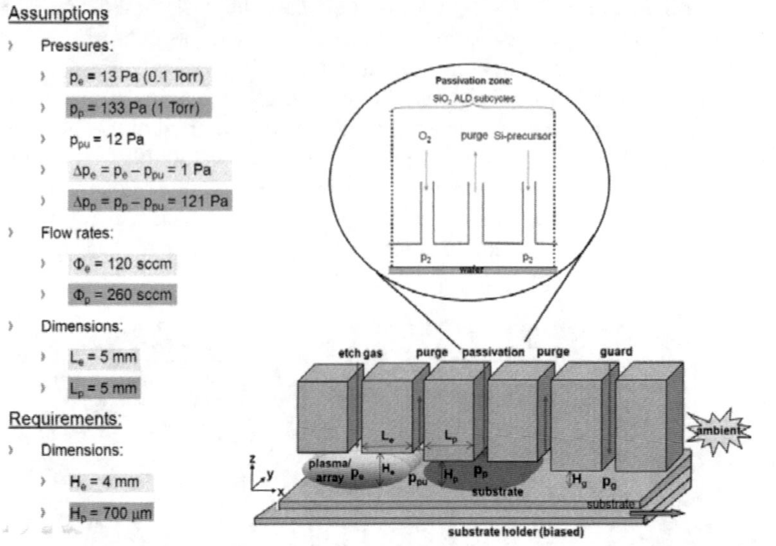

Figure 4. Schematic of the multichannel gas injector head in spatial DRIE. The injector contains the etch gas inlet, the passivation gas inlet(s) and an outer guard gas inlet to screen off the ambient atmosphere. All inlets can be separated from each other by purge outlets (drawn) and by bearing gas inlets (not drawn here, but see Fig. 2c). Each gas is injected through an inlet with or without microplasma source.

Thus the energy of the fluorine ions from the etching plasma will be higher than that for any ions attracted from the passivation plasma (more collisional losses for F-ions in case of C_4F_8 passivation or oxygen ions in case of oxide ALD passivation). Therefore, in an entirely spatial process with continuous voltage biasing of the full substrate, the ion bombardment of the passivation layer will be sustained during the etch half-cycle.

Timescale Analysis for Convection, Diffusion, Deposition and Mass Supply

In order to further optimize the spatial SF_6/O_2 etch and SiO_2 deposition process parameters preliminary gas transport simulations were performed to analyze all relevant timescales involved. The simulation program used is a general purpose CFD model CVD-X developed to predict and optimize deposition processes in the semiconductor industry [16]. In this program specific models for the description of rarefied gas transport inside trenches have been incorporated. Using these models, transient multi-scale simulations have been performed of flow, precursor transport and deposition reactions in ALD-type reactors filled with high aspect ratio trenched wafers.

A short synopsis of the most relevant formulas involved is given in Table 1. For more details on the simulation program one is referred to ref. 16. The simulations were done for the passivation of a wide range of lateral feature scales. The three main categories of features studied are a) microsystem cavities with 50 µm openings and aspect ratio 5:1, b) 1 µm wide 3D-vias (aspect ratio 10:1, areal density 100/mm^2) and c) sub-micron (0.15 µm) trenches (aspect ratio 10:1, areal density 10^4/mm^2); see Table 2.

Figure 5 shows some of the main simulations results. For a typical conventional DRIE reactor with ~30 liters volume and process parameter (flow rates, pressure, temperature) settings of ref. 12 the relevant process time is dominated by the flushing time scale, which is of the order of seconds (6.12 sec for 90 % volume flushed; 12.25 sec for 99 % volume flushed). The calculated optimum saturation time (transition point from the Langmuir-dominated regime to the supply-dominated regime) at 50 °C is around 20-50 ms, as shown in Figure 5a. This corresponds very well with the experimentally determined Si-precursor dosing times for saturation [12].

The results in miniaturized reaction zones of 2.5 mm height and 5 mm length (in both directions), representative for our spatial reactor dimensions, indicate an optimal pressure range from ~0.5 Torr for flat structures (Fig. 5a) to ~5 torr for TSVs (Fig. 5b) and microsystems (Fig. 5c), at ~100 °C, and 5 - 200 ms timescales depending on the feature's aspect ratio; see Figs. 5b-d.

It is evident that for lower temperatures and higher aspect ratio features the timescales increase (more Langmuir-dominated). Yet, the corresponding timescales for depositing a few passivation monolayers remain typically in the sub-second regime, even at room temperature, i.e. the targeted process temperature for DRIE in the spatial regime.

Considering the mass supply needed to etch bulk silicon (specific density of $5 \cdot 10^{22}$ atoms/cm^3) it will be obvious that the Si-etching half-cycle requires prolonged time intervals. This requirement is the main driving force for the development of high-density plasmas in Si-etching.

Table 1. Formulas used for the timescale analysis of convection, diffusion, deposition and mass supply in the different 3D feature cases, listed in Table 2. (a = aspect ratio; A = area). More details in ref. [16].

Eq. 1. Convection, flushing reactor volume	$\tau_{supply,gas} = \dfrac{P_{tot} \cdot M_{precursor} \cdot V_{reactor}}{R \cdot \Phi_{tot}}$	
Eq. 2. Reactor diffusion	$\tau_{diff,reactor} = \dfrac{H^2_{reactor}}{D}$	
Eq. 3. Trench diffusion	$\tau_{diff,trench} = \dfrac{L^2_{trench}}{D_{Kn}}$	$D_{Kn} = \dfrac{4D_{trench}}{3} \cdot C_0$
Eq. 4. Langmuir deposition flat	$\tau_{Langmuir,flat} = \dfrac{n''_{tot} \cdot R \cdot T}{c_{gs} \cdot f_s \cdot C_0 \cdot P_{precursor}}$	$C_0 = \sqrt{\dfrac{R \cdot T}{2\pi \cdot M_{precursor}}}$
Eq. 5. Langmuir deposition trenches	$\tau_{Langmuir,trench} = \tau_{Langmuir,flat} \cdot \left(\dfrac{3}{2}a^2 + \dfrac{19}{4}a + 1\right)$	
Eq. 6. Required mass saturating flat	$\tau_{supply,flat} = \dfrac{A_{flat} \cdot d_{layer} \cdot \rho_{solid}}{c_{gs} \cdot (M_{solid}/M_{precursor}) \cdot \Phi_{precursor}}$	
Eq. 7. Required mass saturating trenches	$\tau_{supply,trench} = \dfrac{A_{trench} \cdot d_{layer} \cdot \rho_{solid}}{c_{gs} \cdot (M_{solid}/M_{precursor}) \cdot \Phi_{precursor}}$	

Table 2. Dimensions and densities of three characteristic 3D-features used in the timescale analysis of convection, diffusion, deposition and mass supply in spatial DRIE: microsystem cavities with 50 μm openings and aspect ratio 5:1; 1 μm wide 3D-vias (aspect ratio 10:1, areal density 100/mm^2) and sub-micron (0.15 μm, DRAM-like) trenches (aspect ratio 10:1, areal density 10^4/mm^2). Flat wafers are used as a reference.

Feature characteristic	Flat wafer	μ-systems & Sensors/actuators	TSVs	DRAM
Type	-	MEMS cavities	TSV	trench (pore)
Diameter/width (μm)	0	50	1	0.15
Depth (μm)	0	250	10	1.5
Aspect ratio	-	5	10	10
Density (number/mm^2)	0	10	100	10^4
Exposed area (%)	0	2	0.008	0.02
Area multiplier	1.0	1.39	1.003	1.007

Microplasma sources

The dimensions of the DRIE reactor design described above, call for the use of miniaturized plasma sources or arrays. In view of the accelerated etch rate requirements a logical further step is to make the etch cycle proceed at higher pressure, ideally at atmospheric pressure, or at least sub-atmospheric, *e.g.* 100 mTorr. At higher pressures one can expect higher electron densities (n_e), and correspondingly higher ion and radical formation [17]. Power densities achieved in microdischarges (kW.cm^{-3} to MW.cm^{-3}) are orders of magnitude larger than those in conventional large-scale systems (W.cm^{-3}) [18].

Today, high-density plasma sources are now being designed and explored by the use of microplasma sources or arrays. Generally, this research aims at achieving a ~hundred-fold increase in electron density [19], significantly beyond the traditional densities of 10^{12} cm^{-3}, thus enabling high-speed etching at correspondingly higher pressures. The challenge is to avoid increased ion scattering, so that the energy and directionality of the ion bombardment is maintained to combine increased etch rates with good anisotropy control in 3D Si etching. Figure 6 shows the potential gain in plasma density (n_e) upon miniaturization of ultrahigh-frequency plasma sources in combination with the use photo-stimulation by UV light sources.

The new concept of ALD passivation for RIE in an all-spatial regime can also be combined with microplasma sources replacing the traditional plasma sources.

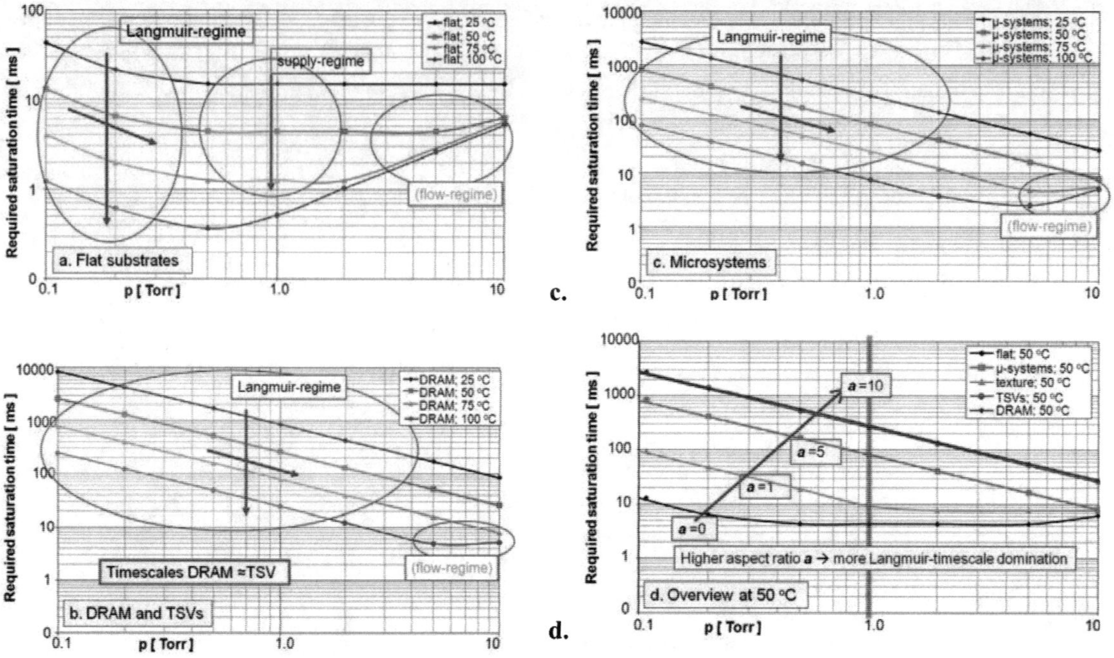

Figure 5. Estimated timescales for a) flat substrates and for various categories of 3D features with different densities (*cf.* Table 2): b) 0.15-1 µm 3D trenches and vias with aspect ratio 10:1, c) microsystems with 50 µm trenches with aspect ratio 5:1, and d) overview at 50 °C.

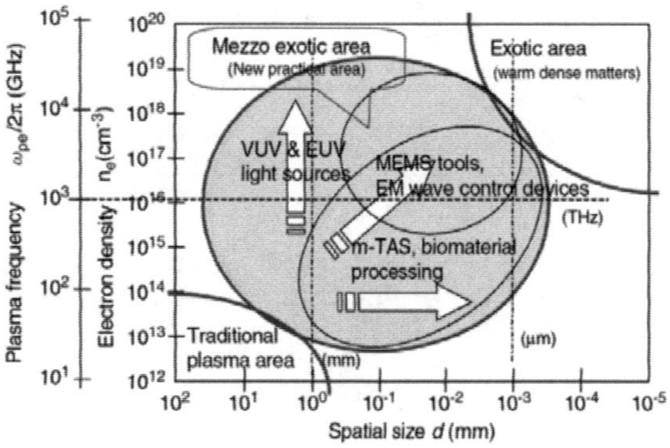

Figure 6. Characteristics of microplasmas in a plane of the spatial size d and the plasma density n_e. From ref. [19]; © IEE Japan, 2006.

As a first step we have designed a Dielectric Barrier Discharge microplasma source with a rectangular (0.5 mm wide x 5 mm long) opening for the reactant gas, which is now being tested and optimized for its etching behavior at close distances from an RF-biased substrate (≤ 2 cm). The etching is done in combination with passivation by C_4F_8 CVD and by ALD-based oxide (Al_2O_3). Preliminary results obtained are not yet optimized: the maximum Si-etch rate achieved so far was 4 µm/min at only 35 W plasma source power, 10 V substrate bias, 1.2 mbar, 50 sccm SF_6 and 2 cm distance between the microplasma source and the Si-sample. We are confident that further optimization (higher powers and concentrations, shorter distances of ≤ 0.5 cm, SiO_2 ALD passivation, and light stimulation) will give the necessary improvement.

Concluding remarks and outlook

We have described a novel alternative and disruptive concept of Deep Reactive Ion Etching, which converts the conventional time-divided process into the *spatially-divided* regime. The spatial separation is realized by inert gas bearing 'curtains' of heights above the wafers down to ~20 µm. These curtains confine the reactant gases to individual (e.g. linear) injection slots constructed in a gas injector head. By horizontally moving the substrate back and forth under the head one can expose the wafer to the alternate gases in the overall cycle without the intermittent reactor volume refreshment time delay that occurs upon every cycle.

Spatially-divided Deep Reactive Ion Etching has the potential of yielding unprecedented etch rates in 3D Si etching and becoming a future high-speed alternative for the conventional Bosch process in cost-effective creation of advanced 3D interconnects (TSVs), MEMS manufacturing and other applications.

An additional advantage of the spatial DRIE regime is the significant reduction of passivation that settles on (and flakes from) the reactor walls since in this regime no etch or passivation products deposit. Moreover, fluorine-free, thus environmentally friendlier passivation chemicals can be used. When, in addition, the etch process is applied with ALD-based SiO_2 passivation rather than with CVD-based C_4F_8-based half-cycles, the anisotropy control can be simplified, and yield *e.g.* reduced scallops and mask undercut.

Finally, we note that the concept of spatial etching can also be applied in other demanding application fields such as (plasma enhanced) ALE, *i.e.* Atomic Layer Etching for nanoscale MOSFET devices requiring etching with atomic layer resolution [20,21] and supreme sidewall passivation to suppress line edge roughness. This technology, based on cycles similarly composed of etch/purge/passivation/purge is not commercially viable yet, but can certainly be optimized in the spatially divided regime.

Acknowledgement
The authors wish to thank S. Riedel, M. Mildner, J. Sundqvist (Fraunhofer CNT, Dresden) and E. Erben (Namlab, Dresden) for providing the ultradeep trenched wafers that we used for ALD deposition (Fig. 3b). Prof. D.C. Schram (Eindhoven University of Technology) is acknowledged for in-depth discussions on microplasma source design.

References

[1] Shockley W US Patent 3,044,909, filed 23 Oct, 1958; granted 17 July, 1962
[2] Smith M and Stern E US Patent 3,343,256, filed 28 Dec., 1964; granted 26 Sept., 1967
[3] *Handbook of 3-D Integration: Technology and Applications of 3D Integrated Circuits* 2008, ed Garrou P, Bower C, and Ramm P (Weinheim: Wiley-VCH Verlag)
[4] Lärmer F and Schilp A, US Patent 5,498,312, 12 March, 1996;
[5] Lärmer F, *et al.*, 2010 *Handbook of Silicon Based MEMS Materials and Technologies*, ed V Lindroos *et al.*, (Elsevier: Oxford) p 349
[6] Wu B, Kumar A and Pamarthy S 2010 *J. Appl. Phys.* **108** 051101
[7] Roozeboom F, Lankhorst A M, Poodt P W G, N.B. Koster, Winands G J J and Vermeer A J P M, '*Apparatus and method for reactive ion etching*', Patent application WO 2011/105908, 1 Sept. 2011
[8] Boufnichel M, Lefaucheux P, Aachboun S, Dussart R and Ranson P 2005 *Microelectron. Eng.* **77** 327
[9] Tegen S and Moll P, 2005 *J. Electrochem. Soc.* **152** G271
[10] Sainiemi L and Franssila S, 2007 *J. Vac. Sci. Technol.* B **25** 801
[11] Kobrin B, '*Method and apparatus for anisotropic etching*', Patent application US2010173494 (A1), 8 July 2010
[12] Dingemans G, van Helvoirt C A A, Pierreux D, Keuning W and Kessels W M M, 2012 *J. Electrochem. Soc.* **159** H277
[13] Knoops H C M, Langereis E, van de Sanden M C M and Kessels W M M, 2010 *J. Electrochem. Soc.* **157** G241
[14] Poodt P, Lankhorst A, Roozeboom F, Spee C, Maas D and Vermeer A, 2010 *Adv. Mater.* **22** 3564, and references therein
[15] Poodt P, Cameron D C, Dickey E, George S M, Kuznetsov V, Parsons G N, Roozeboom F, Sundaram G and Vermeer A, 2012 *J. Vac. Sc. Technol.* A **30** 010802-1
[16] Lankhorst A M, Paarhuis B D, Terhorst H J C M, Simons P J P M, Kleijn C R, 2007 *Surface and Coatings Technology* **201**, 8842
[17] Schram D C, 2009 *Plasma Sources Sci. Technol.* **18** 014003
[18] Iza F, Kim G J, Lee S M, Lee J K, Walsh J L, Yuantao T. Zhang Y T, Kong M G, 2008 *Plasma Process. Polym.* **5** 322
[19] Tachibana K, 2006 *IEEJ Trans.* **1**, 145
[20] Agarwal A, Kushner M J, *J. Vac. Sci. Technol. A* 2009 **27**, 37
[21] Park J B, Lim W S, Park B J, Park, I H Kim Y W, Yeom G. Y, 2009 *J. Phys. D: Appl. Phys.* **42**, 055202

IOP Conf. Series: Materials Science and Engineering **41** (2012) 012002 doi:10.1088/1757-899X/41/1/012002

Ge technology beyond Si CMOS

Albert Chin
Dept. of Electronics Engineering, National Chiao Tung University, Hsinchu, Taiwan ROC

E-mail: albert_achin@hotmail.com

Abstract. To save energy, low voltage operation is the most important criterion for CMOS ICs. To reach this goal, high mobility new channel materials are required for CMOS ICs at ≤14 nm technology nodes. The high electron mobility InGaAs nMOSFET and high hole mobility Ge pMOSFET were proposed for CMOS at 0.5 V operation, since the poor hole mobility of InGaAs makes it unsuitable for all InGaAs CMOS. However, the epitaxial InGaAs nMOSFET on Si faces fundamental material challenges with large defects and high leakage current. Although dislocation-defects-free Ge-on-Insulator (GeOI), ultra-thin-body (UTB) InGaAs IIIV-on-Insulator (IIIVOI), and selective GeOI on Si were pioneered by us, it is still difficult to reach InGaAs-nMOS/Ge-pMOS CMOS targeting to ≤14 nm CMOS. In contrast, Ge is the ideal candidate for all Ge CMOS logic due to both higher electron and hole mobility than Si. Significantly higher (2.6X) hole mobility of GeOI pMOSFET than universal SiO_2/Si value was reached at a medium 0.5 MV/cm effective electric field (E_{eff}) and 1.4 nm equivalent-oxide-thickness (EOT). Nevertheless, the Ge nMOSFET suffers from large EOT and fast mobility degradation with increasing E_{eff}, due to the surface Fermi-level pinning to valance band, poor high-κ/Ge interface and low dopant activation. Using novel laser annealing and proper gate stack, small EOT of 0.95 nm, small sub-threshold swing of 106 mV/dec, and 40% better high-field mobility than universal SiO_2/Si data were achieved in Ge nMOSFET. Such all-Ge CMOS has irreplaceable merits of much simpler process, lower cost, and potentially higher yield than the InGaAs-nMOS/Ge-pMOS CMOS platform.

1. Introduction

The IC chips consume a large portion of energy globally and will increase more in the near future. Therefore, low voltage CMOS must be developed to lower the energy and power consumption, according to fundamental physics of $P=C_{load}V_D^2f$, where the C_{load}, V_D, and f are the load capacitance, driven voltage, and operation frequency of an IC. To maintain high performance operation at a low voltage, high mobility new channel materials are necessary for CMOS IC at ≤14 nm technology nodes. Thus, high electron mobility InGaAs nMOSFET and high hole mobility Ge pMOSFET were proposed for ≤14 nm CMOS at 0.5 V operation, which can lower the AC switching power as much as 50% from the current 0.7 V V_D bias used for 22 nm CMOS ICs. However, the integration of InGaAs on Si faces fundamental material challenges: the large 8% lattice-mismatch, high dislocation densities, and anti-phase domain boundaries by polar-nonpolar mismatched lattice. These defects further lead to high leakage current in epitaxial grown InGaAs nMOSFET on Si substrate. Although the device improves with increasing the buffer layer thickness, this will cause the integration of InGaAs nMOSFET with bottom Si devices more difficult.

To address these epitaxial growth issues, dislocation-defects free Ge and InGaAs on Si were demonstrated by us using the Ge-on-Insulator (GeOI) [1]-[2], ultra-thin-body (UTB) InGaAs IIIV-on-Insulator (IIIVOI) [3]-[4], and selective GeOI on Si [5] using the wafer bonding, smart-cut, or selective etching method. Alternatively, Ge has higher electron (3X) and hole (4X) mobility than Si that is the ideal candidate for all Ge CMOS logic. This is quite different to the InGaAs case, where the poor hole mobility prevents to form all InGaAs CMOS. This all-Ge CMOS approach is much simpler than the CMOS platform of InGaAs nMOSFET and Ge pMOSFET, which has unique merits of smaller numbers of mark, simpler process step, lower cost, and potentially higher yield. However, the Ge nMOSFET suffers from large equivalent-oxide-thickness (EOT) and fast mobility degradation with increasing oxide effective electric field (E_{eff}), which are related to the surface Fermi-level pinning to valance band, poor high-κ/Ge interface and low dopant activation. In this paper, we will discuss the potential to use GeOI, IIIVOI, selective GeOI, and the progress of all Ge CMOS.

2. Experiments

The Ge/SiO_2/Si GeOI was formed by first depositing SiO_2 on both Ge and Si wafers using PECVD, O_2 plasma treatment to top SiO_2, bonding Ge/SiO_2 and SiO_2/Si at 500°C, etching back and polishing the Ge [1]. Figure 1 show the fabricated GeOI without and with O_2 plasma treatment. Void-free Ge/SiO_2-SiO_2/Si interface with improved mechanical strength was reached by using O_2 plasma treatment at such low temperature. Dislocation-free Ge is obtained and the smooth Ge/SiO_2 interface allows further thinning down the body thickness.

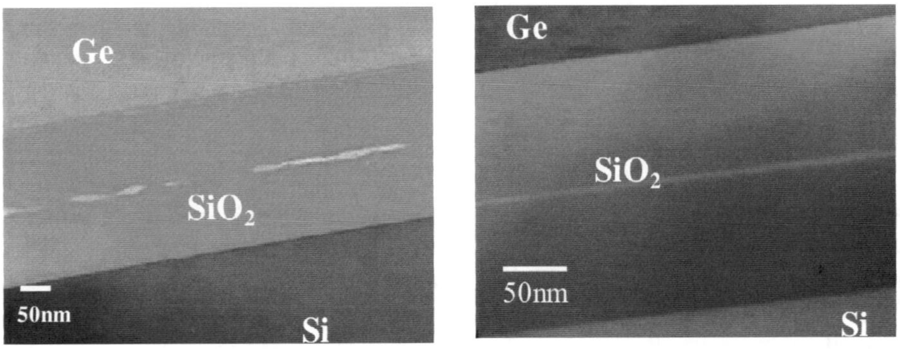

Figure 1. Cross-sectional TEM of low-temperature bonded GeOI without (left) and with (right) O_2 plasma treatment before bonding.

We further fabricated the selective GeOI [5]. As shown in Figure 2, the selective GeOI was formed by SiO_2 deposition on both H^+-implanted Ge and standard 1-Poly Si-6-Metal (1P6M) 0.18μm MOSFETs, O_2 plasma enhanced bonding, smart-cut, followed by 400°C annealing and slight polishing.

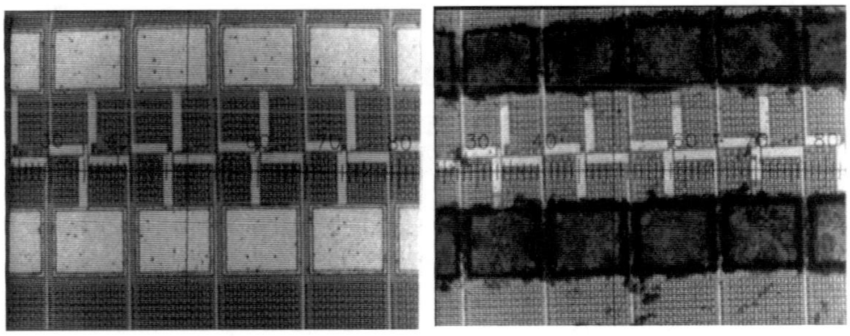

Figure 2. Top view of selective GeOI on 1P6M 0.18μm CMOS before (left) and after (right) bonding.

We also extended the GeOI to wider semiconductor-on-Insulator (SemiOI). Figure 3 shows the fabricated InAlAs/InGaAs IIIVOI [3]-[4]. An inverted InAlAs/InGaAs/InAlAs HEMT hetero-structure was first grown on InP substrate, followed by PECVD SiO_2 deposition, applying O_2 plasma treatment to top SiO_2/InAlAs/InGaAs/InAlAs/n^+-InGaAs/InAlAs/InP and SiO_2/Si, bonding at 400°C, and selective etching the InP substrate and InAlAs buffer layer. Such low temperature bonding process was developed to decrease the thermal mismatch among III-V, SiO_2 and Si materials.

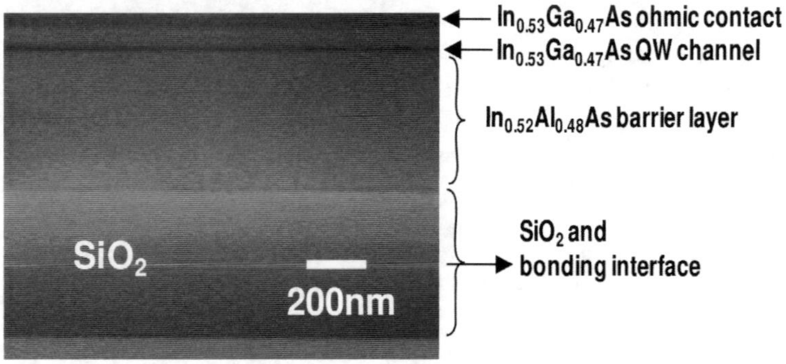

Figure 3. Cross-sectional TEM of InAlAs/InGaAs/InAlAs FET structure.

After GeOI or selective GeOI fabrication, the pMOSFET were made by high-κ $LaAlO_3$ gate dielectric and high work-function IrO_2 metal-gate formation, self-aligned source-drain B^+ implantation, and dopant activation at 500°C. The InAlAs/InGaAs IIIVOI FET was formed by gate-recess etching the top n^+ InGaAs contact layer, Ti/Au Schottky gate lift-off, and NiGeAu source-drain contacts [3].

For high performance Ge nMOSFET, a gate-last process was used. To improve the high-κ/Ge interface, an ultra-thin interfacial SiO_2 [6] and KrF laser annealing (248 nm, ~30 ns pulse) [7]-[8] were applied. The laser annealing was also used to increase the dopant activation of P^+-implanted source-drain. The gate dielectric is composited with ~0.8 nm SiO_2, 1 nm La_2O_3 and 3 nm ZrO_2, followed by 400°C furnace annealing. Then ultra-fast laser annealing was applied to prevent the Ge out-diffusion into high-κ gate dielectric [6]. The Ge nMOSFETs were made by patterning the deposited TaN gate and Al source-drain contact metal.

3. Results and discussion

3.1. Defect-free SemiOI transistors

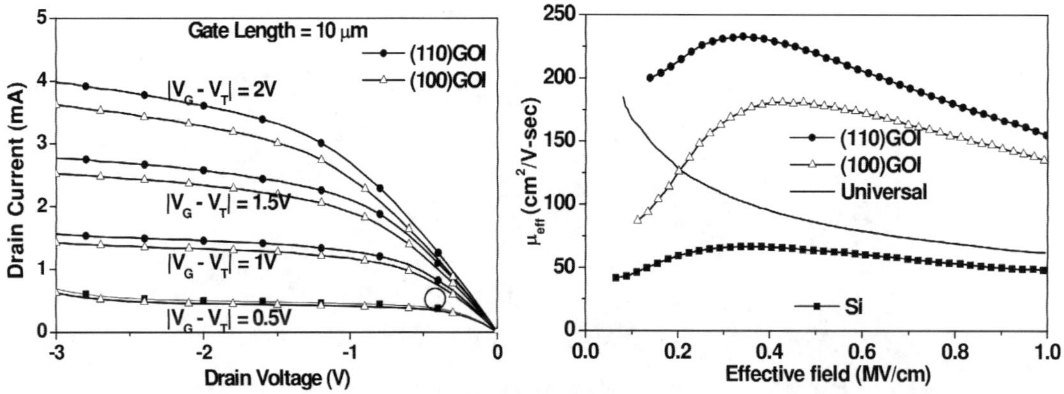

Figure 4. I_d-V_d (left) and mobility-E_{eff} (right) characteristics of IrO_2/$LaAlO_3$/GeOI pMOSFETs.

Figure 4 shows the I_d-V_d and mobility data of $IrO_2/LaAlO_3$/GeOI pMOSFETs. In addition to the good transistor characteristics, the metal-gate/high-κ/GeOI pMOSFET at (110)-orientation has 2.6X higher hole mobility than universal SiO_2/Si mobility, at a medium 0.5 MV/cm E_{eff} and a 1.4 nm EOT. The bonding top layer with different orientation to bottom substrate is the advantage of SemiOI technology.

The device I_d -V_d and gain-frequency characteristics of InAlAs/InGaAs IIIVOI nFETs are shown in Figure 5. A high drive current of 0.41 mA/μm was obtained for this nFET with a 1.1 μm gate length. The good device performance is evident from the high RF current gain ($|H_{21}|^2$) and cut-off-frequency (f_t) of 21 GHz that are close to those of a 0.35μm Si MOSFET with much smaller gate length. The better transistor performance is due to the significantly higher mobility of 8,100cm^2/Vs than that for a Si MOSFET.

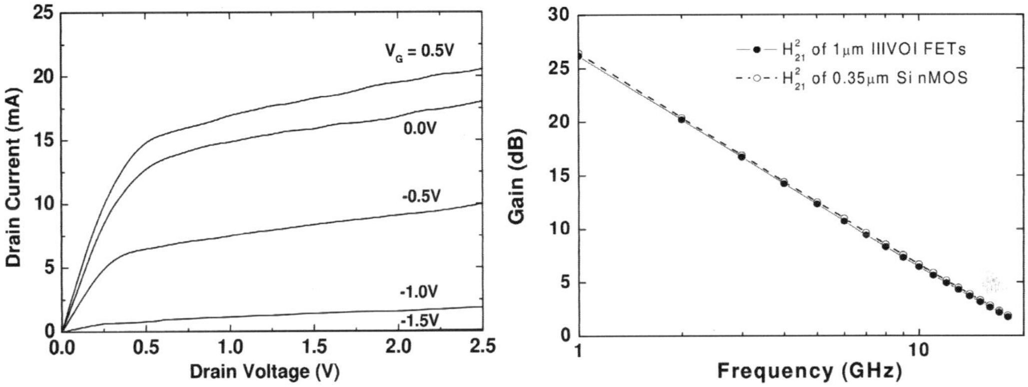

Figure 5. I_d-V_d (left) and gain-frequency (right) characteristics of Au/Ti/InAlAs/InGaAs/IIIVOI nFETs.

It is important to notice that although the selective wafer-bonding on 50-μm×50-μm area was obtained, this technique is still different to reach both the IIIVOI and GOI side-by-side for high mobility InGaAs-nMOS/Ge-pMOS CMOS platform. The low hole mobility makes InGaAs unsuitable for all-InGaAs CMOS. The epitaxial InGaAs nMOSFET on Si is still quite challenging due to the large defects density and high transistor leakage current.

3.2. High mobility Ge nMOSFET for all-Ge CMOS

The Ge has both higher electron and hole mobility than Si - an ideal candidate for all Ge CMOS logic. The GeOI pMOSFET shown in Figure 4 has much higher hole mobility than universal SiO_2/Si mobility at a small 1.4 nm EOT. However, Ge nMOSFET suffers from poor device performance due to surface Fermi-level pinning to valance band, low dopant activation, and poor high-κ/Ge interface. To address these issues, novel laser annealing with fast ~30 ns pulse was used [7]-[9]. The laser annealing melts surface semiconductor and re-crystallize within a very short time, which can fully activate the ion-implanted dopants and form ultra-shallow junction. Such laser annealing was proposed for highly scaled ≤14 nm CMOS to reach both ultra-shallow junction and low series resistance. We pioneered the use of laser annealing on high-κ gate dielectric [7]. Figure 6 shows the device characteristics of $TaN/ZrO_2/La_2O_3/SiO_2$/Ge MOS capacitors and nMOSFETs. Applying laser annealing on gate dielectric not only improves the C-V hysteresis but also increases the gate capacitance. The higher gate capacitance is attributed to the fast ~30 ns annealing with less high-κ/Ge interface reaction, which gives a small EOT of 0.95 nm from quantum-mechanical C-V simulation. From the measured transistor I_d-V_g characteristics, a small sub-threshold swing of 106 mV/dec was obtained that is consistent with the improved C-V hysteresis. Figure 7 shows the mobility-E_{eff} characteristics, after gate leakage correction [10]. At 1 MV/cm E_{eff}, 40% better high-field mobility than universal SiO_2/Si data was achieved in Ge nMOSFET. These results suggest the fast electron mobility degradation is not intrinsic to Ge nMOSFET.

IOP Conf. Series: Materials Science and Engineering **41** (2012) 012002 doi:10.1088/1757-899X/41/1/012002

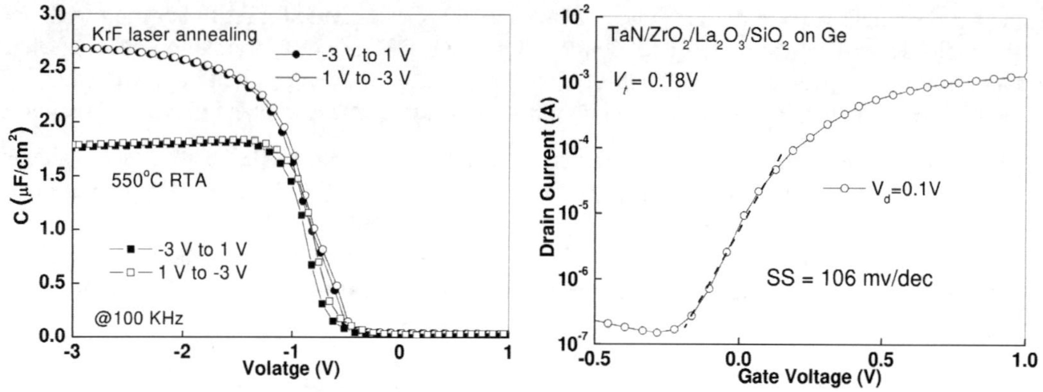

Figure 6. Capacitor C-V (left) and transistor I_d-V_g (right) characteristics of $TaN/ZrO_2/La_2O_3/SiO_2/Ge$ nMOSFETs by laser annealing.

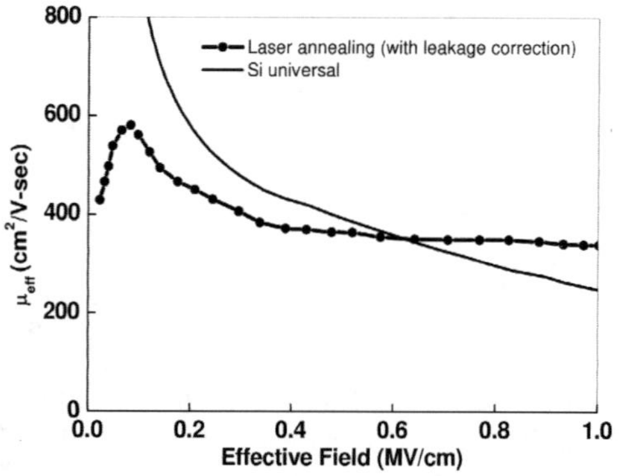

Figure 7. Mobility-E_{eff} characteristics of $TaN/ZrO_2/La_2O_3/SiO_2/Ge$ nMOSFET by laser annealing.

4. Conclusions

High mobility new channel materials are required at ≤14 nm CMOS, but the InGaAs/Si nMOSFET is still very challenging due to fundamental material issues with large defects and high leakage current. Although dislocation-defects-free InGaAs-IIIVOI, GeOI, and selective GeOI were developed, it is still difficult to reach InGaAs-nMOS/Ge-pMOS CMOS side-by-side with nm-scale dimension for ≤14 nm CMOS. In contrast, the GeOI pMOSFET shows 2.6X higher hole mobility than universal SiO_2/Si data at a medium 0.5 MV/cm E_{eff} and 1.4 nm EOT. Using novel laser annealing and proper gate stack on Ge nMOSFET, small 0.95 nm EOT and 40% better 1-MV/cm mobility than universal SiO_2/Si data were achieved. Although further device performance improvement on Ge nMOSFET is required, the fast mobility degradation with increasing E_{eff} is not an intrinsic limitation to Ge nMOSFET. The all-Ge CMOS has irreplaceable merits of much simpler process, lower cost, and potentially higher yield than the InGaAs-nMOS/Ge-pMOS CMOS platform.

Acknowledgments

Albert Chin would like to thank Dr. Yasushi Akasaka, Jim Hu, and Tsunetoshi Arikado for helpful discussion, and the supports from National Science Council, Taiwan and Tokyo Electron Ltd.

IOP Conf. Series: Materials Science and Engineering **41** (2012) 012002 doi:10.1088/1757-899X/41/1/012002

5. References

[1] Huang C H, Yang M Y, Chin A, Chen W J, Zhu C X, Cho B J, Li M F and Kwong D L 2003 Very low defects and high performance Ge-On-Insulator p-MOSFETs with Al_2O_3 gate dielectrics *Symp. on VLSI Tech. Dig.* pp 119-120

[2] Huang C H, Yu D S, Chin A, Chen W J, Zhu C X, Li M F, Cho B J, and Kwong D L 2003 Fully silicided NiSi and germanided NiGe dual gates on SiO_2/Si and Al_2O_3/Ge-on-Insulator MOSFETs," *Int. Electron Devices Meeting Tech. Dig.* pp 319-322

[3] Chin A, Chen C, Yu D S, Kao H L, McAlister S P and Chi C C 2006 Comparison of Germanium-on-Insulator CMOS with InGaAs MOSFETs *Materials Science in Semiconductor Processing (European Materials Research Society)* vol 9 pp 711-715

[4] Liao C C, Kao S, Chin A, Yu D S, Li M F, Zhu C, and McAlister S P, 2006 Comparing high mobility InGaAs FETs with Si and GOI devices," *64th Device Research Conference* pp 85-86

[5] Yu D S, Chin A, Laio C C, Lee C F, Cheng C F, Chen W J, Zhu C, Li M F, McAlister S P and Kwong D L 2004 3D GOI CMOSFETs with novel IrO_2(Hf) dual gates and high-κ dielectric on 1P6M-0.18μm-CMOS *Int. Electron Devices Meeting Tech. Dig.* pp 181-184

[6] Chen W B and Chin A 2009 Interfacial layer dependence on device property of high-κ TiLaO Ge/Si *n*-type metal-oxide-semiconductor capacitors at small equivalent-oxide thickness *Appl. Phys. Lett.* **95** 212105

[7] Chen W B, Shie B S, Cheng C H, Hsu K C, Chi C C and Chin A 2010 Higher κ metal-gate/high-κ/Ge n-MOSFETs with <1 nm EOT using laser annealing *Int. Electron Devices Meeting Tech. Dig.* pp 420-423

[8] Liao C C, Chin A, Su N C, Li M F and Wang S J, 2008 Low V_t gate-first Al/TaN/[Ir_3Si-$HfSi_{2-x}$]/HfLaON CMOS using simple process *Symp. on VLSI Tech. Dig.* pp 190-191

[9] Chen W B, Shie B S and Chin A 2011 Higher gate capacitance Ge n-MOSFETs using laser annealing *IEEE Electron Devices Lett.* **32** 449

[10] Zhu W J, Han J P and Ma T P 2004 Mobility measurement and degradation mechanisms of MOSFETs made with ultrathin high-κ dielectrics *IEEE Trans. Electron Devices* **51** 98

GaAs clean up studied with synchrotron radiation photoemission

Massimo Tallarida[1,3], Christoph Adelmann[2], Annelies Delabie[2], Sven van Elshocht[2], Matty Caymax[2], and Dieter Schmeisser[1]

[1] Brandenburg University of Technology, Applied Physics – Sensors, Konrad-Wachsmann-Allee, 17, 03046, Cottbus
[2] Imec, Kapeldreef 75, B-3001 Leuven, Belgium

[3] E-mail: tallamas@tu-cottbus.de

Abstract. In this contribution we describe the chemical changes at the surface of GaAs upon adsorption of tri-methyl-aluminum (TMA). TMA is used to grow Al_2O_3 with atomic layer deposition (ALD) usually using H_2O as oxygen source. Recently, it was pointed out that the adsorption of TMA on various III-V surfaces reduces the native oxide, allowing the growth of an abrupt III-V/High-K interface with reduced density of defects. Synchrotron radiation photoemission spectroscopy (SR-PES) is a powerful method to characterize surfaces and interfaces of many materials, as it is capable to determine their chemical composition as well as the electronic properties. We performed in-situ SR-PES measurements after exposing a GaAs surface to TMA pulses at about 250°C. Upon using the possibility of tuning the incident photon energy we compared the Ga3d spectra at 41 eV, 71 eV, 91 eV and 121 eV, as well as the As3d at 71 eV and 91 eV. Finally, we show that using SR-PES allows a further understanding of the surface composition, which is usually not accessible with other techniques.

1. Introduction

A possible direction that could be trod to follow the "more than Moore" philosophy to go behind the actual use of standard semiconductors/oxides microstructures to form electronic devices, is to use semiconductors with high carrier mobility in combination with oxides with large dielectric constants (high-k) [1]. Compound semiconductors formed by elements of the III and V group have usually very high electron mobility because of the large curvature of the conduction band next to the gamma point in the reciprocal space [2]. Therefore, they represent suitable substitutes to n-Si. An important advantage of using III-V semiconductors over Si (and in general over elemental semiconductors) is represented by the possibility of using various alloy combinations in order to vary both the lattice constant and band gap [3]. In this way, it is clear that III-V semiconductors are versatile materials that could be combined together and used for different purposes on the same device in an heterogeneous system integration [1]. Although III-V semiconductors show these interesting bulk properties, their use is still not fully developed, because of inconvenient surface properties, namely the large number of defects formed in many different ways, ranging from dimer formation to antisite defects, interstitials [4]. All these kind of defects deliver a large density of defect states, which usually pin the Fermi level around mid-gap [4]. In the general understanding of surface defects, their formation is related to the growth of a complex native oxide, which degenerates the underlying layers of the III-V

IOP Conf. Series: Materials Science and Engineering **41** (2012) 012003 doi:10.1088/1757-899X/41/1/012003

semiconductors. Many efforts have been made to find reliable ways to reduce surface defects, e.g. eliminating the low-quality oxide and passivating the defect states with sulfur [4]. Recently, it was observed that upon growing high-k oxides by atomic layer deposition (ALD), the native oxide of many III-V semiconductors could be partially removed (clean-up or self-cleaning process) [5,6,7]. This mechanism could be therefore used to reduce the density of defects and eventually unpin the Fermi level. However, the native oxide removal is usually only partial and the density of defects remains too large to unpin the Fermi level [8]. In order to determine whether the oxide reduction upon ALD can effectively unpin the Fermi level of the underlying semiconductor, it is necessary to understand which kind of defects are producing the Fermi level pinning and which mechanisms control the native oxide reduction. To this aim, a deep knowledge of the actual morphology of the native oxide is necessary. The standard method to eliminate the native oxide of GaAs (maybe the simplest and most investigated III-V semiconductor and therefore a model system) is to use Tri-Methyl-Aluminum (TMA) pulses, while heating at temperatures ranging between 200 °C and 300 °C. We used this model system to determine whether TMA effectively removes the native oxide completely, and to find which species eventually persist or are rather produced after TMA pulses. We conducted in-situ ALD experiments, where samples before after TMA pulses were investigated in an ultra-high-vacuum (UHV) chamber attached to the ALD reactor and separated by a valve [9]. As investigation method we used photoemission spectroscopy since it is a very effective method to study surfaces and interfaces of many materials. This is especially true when photoemission spectroscopy is made using synchrotron radiation as a light source. The reason behind it is the dependency of the electrons escape depth on their kinetic energy. In particular, the minimum electron mean free path is usually considered to find at 50 eV kinetic energy, although this value is not sharp and depends on materials properties, like atomic density [10]. The minimum electron mean free path value is often considered to be around 0.5 nm, although also this value is not always demonstrated experimentally. However, tuning the photon energy around values where the electron kinetic energy is about 50 eV is the best method to determine structural and electronic changes near the surface with photoemission spectroscopy.

We will show photoemission spectra of Ga3d, As3d and O2s peaks at varying photon energies for a GaAs sample prepared with an ex-situ NH4OH dip and after one TMA pulse. The TMA pulse brings strong changes in Ga3d and As3d spectra. These changes are, however, dependent on the photon energy used.

2. Experimental

Photoemission spectra were measured using a Phoibos 150 electron analyzer of SpecsLab GmbH. Photons were delivered by the TMG7 beamline at Bessy II of the Helmholtz Center Berlin. The TGM7 beamline uses a bending magnet to produce the broad synchrotron spectrum that is monochromatically filtered in the range between 15 eV and 125 eV by the toroidal grating monochromator (TGM). The photon resolution depends on the energy and is best at about 50 eV, being about 50 meV, than increasing almost linearly with photon energy reaching about 150 meV at the highest photon energy [8]. The surface sensitivity was increased by measuring photoemission in a 45 ° emission angle geometry.

The ALD reactor is attached to the UHV chamber as described elsewhere [9]. After the TMA pulse of 0.5s and N_2 purges (two purges of 0.5 s separated by 2 seconds), the sample was transferred into the measurement chamber within few minutes. The TMA pulse was performed at about 250 °C.

3. Results and discussion

Surface before the TMA pulse

The Ga3d and As3d spectra of the GaAs sample after the NH_4OH ex-situ cleaning measured at, respectively, 41eV and 71eV are shown in figure 1. Together with the raw data, we also show the lineshape fit with Voigt functions as described elsewhere [8]. As can be observed in figure 1, the surface of GaAs is complex and composed by various chemical species. Gallium is present as Ga^{3+}, forming Ga_2O_3, and as $Ga^{(1+x)+}$, forming a sub-stoichiometric oxide (formally defined as Ga_2O_{1+x}). More complex is the As3d spectrum, where As^{0+} (metallic As) and As-As dimers are present together with the stoichiometric and sub-stoichiometric oxides (As^{3+} and $As^{(1+x)+}$ species).

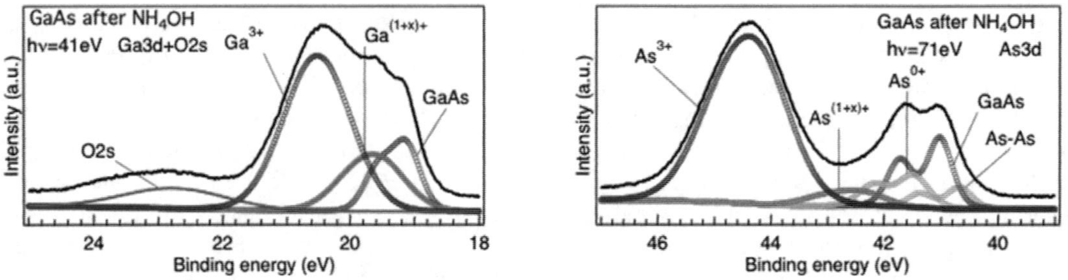

Figure 1: Ga3d+O2s and As3d spectra of the GaAs sample after the NH_4OH ex-situ treatment. Spectra were fitted with Voigt functions as described in reference 8.

The lineshape of the single Voigt functions is different because the spin-orbit splitting of Ga3d is smaller than that of As3d (about 0.44eV vs 0.68eV), so the two spin-orbit components are better resolved in the GaAs, As^{0+} and As-As components of As3d with respect to the GaAs component of Ga3d. For the same reason the stoichiometric oxide As^{3+} is broader than Ga^{3+}, as the two spin-orbit components are not resolved because of the Gaussian broadening of oxides. Moreover, the chemical shifts of the oxide components with respect to GaAs are different for Ga3d and As3d. Ga in GaAs can be, in fact, defined in a formal oxidation state +0.5, while As in -0.5 (considering a charge transfer from Ga to As in a mixed ionic-covalent bond scheme [11]). Since the chemical shift in photoemission depends on the different oxidation state and is smaller for Ga3d (about 0.6eV per oxidation state) than for As3d (about 1eV per oxidation state), the energy difference between Ga3+ and GaAs is only about 1.5eV, while the difference between As3+ and GaAs is about 3.5eV.

The intensity of the various components can be used to infer the morphology of the GaAs after the NH_4OH. If we consider the value of the electron mean free path (λ) to be 0.5nm, the intensity of photoemission depends on depth d as $I=I_0\times\exp(-d/\lambda\cos\vartheta)$, where ϑ is the emission angle (it was fixed to 45° in all experiments shown here) and I_0 is the intensity of the peak without attenuation: it depends on the photoemission cross-section, the number of emitting atoms, the analyzer response function. Upon integrating the signal intensity over the direction perpendicular to the surface, one finds that the total intensity of a peak originated by atoms extending from the top of the surface to a depth d_0 is $I=I_0(1-\exp(-d_0/\lambda\cos\vartheta)$. One can therefore use this relation to determine the extension of the superficial oxide layer, at least within certain conditions, e.g. presence of an abrupt interface between oxide, sub-oxide and pure material; homogeneity of the various layers, independence of λ from the material. In table 1 are listed the peak areas of all components used to fit the As3d peak, while in table 2 the peak areas of the Ga3d fit.

Table 1: Peak area of the five Voigt functions used to fit the spectrum in figure 1b.
Peak Intensity (area x 10^3)

Peak	Intensity (area x 10^3)
As-As	67 ± 3
GaAs	215 ± 9
As^{0+}	121 ± 5
$As^{(1+x)+}$	74 ± 3.5
As^{3+}	870 ± 40

Table 2: Peak area of the three Voigt functions used to fit the spectrum in figure 1a.
Peak Intensity (area x 10^3)

Peak	Intensity (area x 10^3)
GaAs	363 ± 20
$Ga^{(1+x)+}$	410 ± 21
Ga^{3+}	1210 ± 55

We can consider the peaks As-As, As^{0+} and As^{3+} as due to the most superficial atoms, while the $As^{(1+x)+}$ could come from atoms in an intermediate position and finally GaAs due to bulk atoms. In this way the intensities of As-As, As^{0+} and As^{3+} can be added, giving about 1000x103. The ratio between the surface species, the sub-oxide, and GaAs is calculated to be x:0.07x:0.21x. In case of Ga3d there are only three species, which ratio is y:0.34y:0.3y.

Knowing these ratios, however, does not allow the univocal determination of the sample composition near to the surface, and further assumptions should be considered. For example the strong intensity of the As_2O_3 peak with respect to the GaAs in As3d makes reasonable the assumption that the surface termination is given by the As species. Underneath the top As-derived layers the other oxide should be present, namely Ga_2O_3. Moreover, the strong Ga-suboxide compared to the As-suboxide could mean that the former is just underneath Ga_2O_3, while the latter is mostly terminating the bulk GaAs. This layer sequence (model 1) is shown in figure 2 together with the graphic showing the behaviour of the relative and integrated intensity of photoemission peaks as a function of depth.

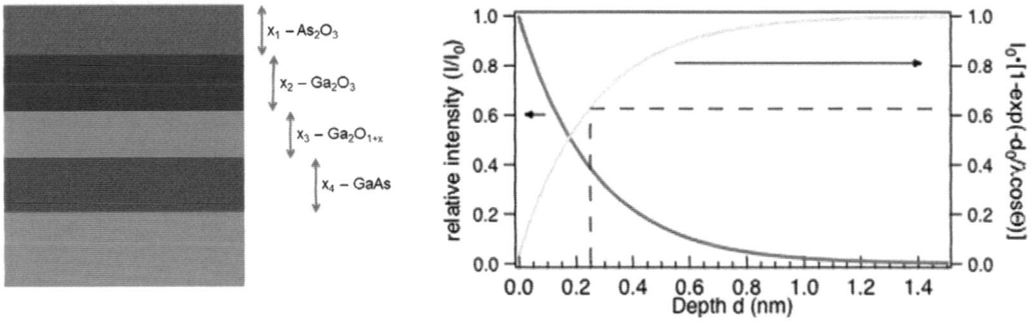

Figure 2: (left) Layer sequence of the major components found in As3d and Ga3d spectra (model 1). (right) Relative intensity (I/I_0) (left axis) and integrated intensity (right axis) of photoemission peaks as a function of depth as discussed in the text. The parameters used were λ=0.5nm and ϑ=45°. Also shown is the example for a signal extending from the surface to a depth of 0.25nm, giving an integrated signal about 60% of the total spectrum intensity.

Although the assumptions leading to the model shown in figure 2 are not validated by neither theoretical calculations nor comparison of intensities from spectra measured with different photon energies, we will use it here to show how it might be used to determine the thicknesses x1 – x4 from the intensities listed in tables 1 and 2.
First of all, the intensities of GaAs components in As3d and Ga3d spectra should be the same since they are originated by electrons belonging to atoms from the same region. In this way the ratios x:0.07x:0.21x and y:0.34y:0.3y can be normalized by imposing 0.21x=0.3y. In this case, and neglecting the small component $As^{(1+x)+}$, we obtain the sequence 1:0.7:0.24:0.21 for, respectively, As_2O_3, Ga_2O_3, Ga_2O_{1+x}, and GaAs. This ratio sequence can be then further normalized to have the total intensity 1, obtaining 0.46:0.33:0.11:0.1. It means: the integrated As_2O_3 signal is 46% of the total detected electrons, Ga_2O_3 the 33%, Ga_2O_{1+x} the 11%, and GaAs the 10%. These values are shown in figure 3. The resulting thicknesses are 0.16nm, 0.26nm, and 0.2nm for, respectively, As_2O_3, Ga_2O_3, and Ga_2O_{1+x} (figure 3, left).

Obviously, the results obtained above, strongly depend on the chosen model. If instead of model 1 we consider a sequence where Ga_2O_3 is on the surface, Ga_2O_{1+x} just below, and As_2O_3 just above GaAs (model 2), the same intensity ratios give a thickness sequence of 0.1nm, 0.04nm, and 0.5nm (shown in figure 3, right). Finally, more complex sequences, where the separation between layers is not well defined and mixed layers are present, would deliver infinite solutions for the same set of intensity values. In order to determine which model better represents the surface morphology, we measured a NH_4OH treated GaAs sample at higher photon energies, i.e. 150eV and 640eV, on another beamline (U49-2/PGM2) at Bessy-II. Those results, not shown here as they will be presented in a forthcoming publication, indicate model 1 as the most probable one.

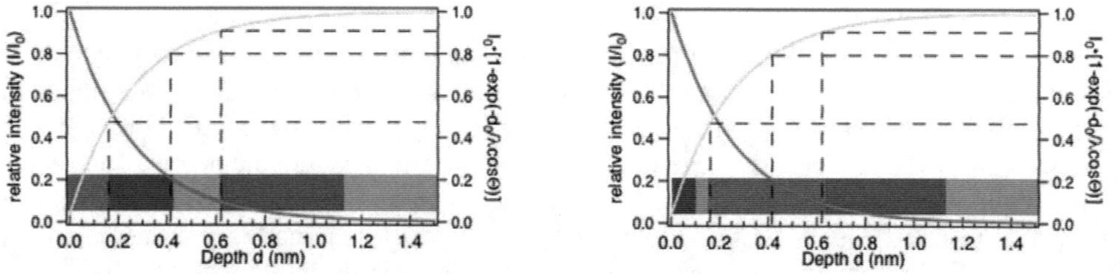

Figure 3: Graphical representation of the thickness calculation of the most important components of As3d and Ga3d spectra. (left) considering the sequence As_2O_3, Ga_2O_3, Ga_2O_{1+x}, and GaAs. (right) considering sequence Ga_2O_3, Ga_2O_{1+x}, As_2O_3, and GaAs.

Surface after 1 TMA pulse

As we showed in a previous publication [8], already one TMA pulse induces strong changes in both Ga3d and As3d spectra. Especially the As^{3+} component of As3d is strongly reduced. This reduction is however accompanied by the growth of the $As^{(1+x)+}$ component, which is related to the sub-stoichiometric As-oxide. The growth of As-suboxide could be due to the reduction of As^{3+} to the less oxidized species, as we previously proposed [8]. In figure 4, we show the comparison of As3d spectra measured at 71eV and 91eV before and after the first TMA pulse. It is evident the strong decrease of the As^{3+} component at both energies, as well as the increase of the GaAs related features. The gallium-related species on the surface behave differently from the arsenic ones. For example, the reduction of the Ga^{3+} species is not so efficient as that of the As^{3+}. However, the amount of Ga_2O_3 is probably larger than that of As_2O_3 as discussed above. Another possible reason could be the different mechanism that induces the Ga_2O_3 reduction. As we showed previously [8], the reduction of Ga_2O_3 is not accompanied by the growth of a sub-stoichiometric oxide, as it is the case for the reduction of As_2O_3. This observation let us consider the possibility that Ga_2O_3 is removed by ligand exchange, i.e. $Al(CH_3)_3(g)+Ga_2O_3(s) \rightarrow Al_2O_3(s)+Ga(CH_3)_3(g)$.

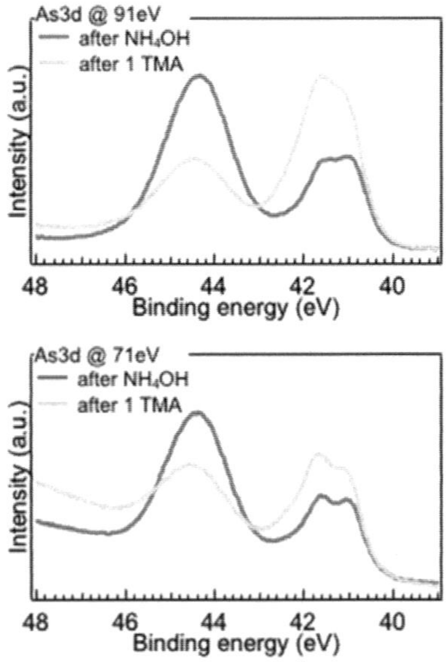

Figure 4: As3d spectra measured at 91eV (top graphic) and at 71eV (bottom graphic) before (red line) and after the first TMA pulse (yellow line).

Another possible reduction mechanism to understand the removal of Ga_2O_3 and correlate it with the growth of As sub-oxide was proposed by Klejna and Elliott [12], where they proposed that reduced arsenic may "getter" oxygen from gallium oxide on the surface, driven by the formation of GaAs, i.e. $1/4Ga_2O_3+As^{+0} \rightarrow 1/2GaAs + 1/4As_2O_3$. This reaction could therefore explain the removal of GaAs if As^{+0} were present on the surface to form GaAs and As-oxide. In general, similar reaction patterns could be possible for reduced As-oxides, if for example the As^{+0} is substituted by $As^{(1+x)+}$. As we found the formation of $As^{(1+x)+}$ after the first TMA pulse and then a continuous decrease similar to that of Ga^{+3} a correlation between the two events could be possible. However, more calculations and experiments should be done in order to clarify whether this mechanism may be taken in consideration. A support in this sense can be given by our measurements of the Ga3d peak at varying photon energies.

This is shown in figure 5, where the measurements of the sample before and after the first TMA pulse are depicted. The energies used were, respectively, 121eV, 91eV, 71eV, and 41eV, starting from the top graphic. The main differences between the spectra before and after the TMA pulse consist in the intensity of the Ga3d peak with respect to the tail on the left side. This is due to the normalization of the curves to have the same intensity of the left side of the spectra. That part is due to the O2s peak. The decrease of Ga3d with respect to O2s means that the distribution of oxygen atoms on the surface is initially dispersed over a large region, which thickness is about 0.6 nm (independently of the model used!). As TMA is pulsed and about 75% of As_2O_3 and 20% of Ga_2O_3 are removed, the oxygen initially bond to As and Ga mostly forms a layer of Al oxide on the surface and oxygen atoms are confined within 0.3nm and, being the intensity of photoemission spectra exponentially dependent on the position of atoms with respect to vacuum, the confinement of oxygen atoms in the first layer brings a stronger peak with respect to the same amount of oxygen atoms dispersed over few layers. The Ga3d peak itself changes very less and maintains the structure with three peaks. Only the ratio is slightly different. This means that after the first TMA pulse, almost all As_2O_3 is removed, oxygen is confined in the first layer and Ga_2O_3 remains almost constant.

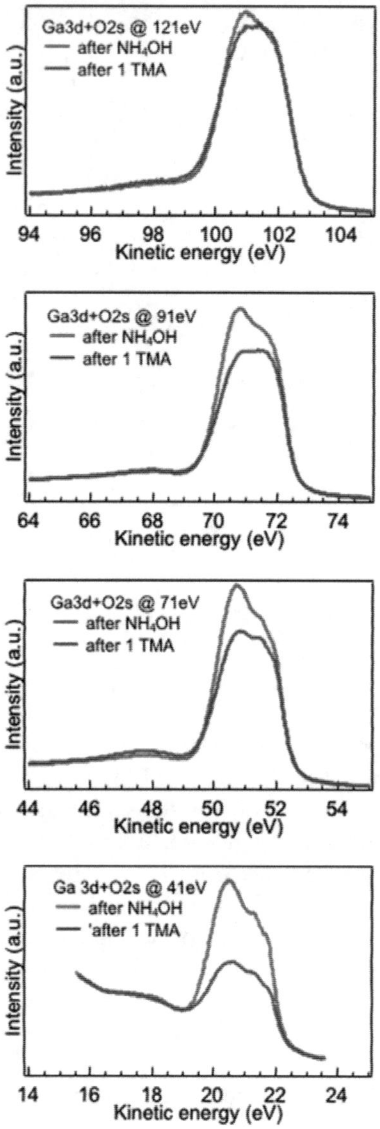

Figure 5: Ga3d and O2s spectra measured at 121eV, 91eV, 71eV, and 41eV before and after one TMA pulse.

4. Conclusions

The structural complexity of GaAs surface makes necessary its study with accurate techniques. In this contribution we showed how synchrotron radiation photoemission may deliver important information about the composition of the surface after the ex-situ chemical treatment (with NH_4OH in this report) and after one TMA pulse. The changes observed are evident and can be discussed in terms of As- and

IOP Conf. Series: Materials Science and Engineering **41** (2012) 012003 doi:10.1088/1757-899X/41/1/012003

Ga-oxide reduction, although with different mechanisms. Further experiments, like photoemission at higher energies, and theoretical calculation may indeed elucidate the composition of the surface and determine whether the changes induced by TMA are capable of reducing the density of defects.

5. Acknowledgements

Authors thank the Bessy-II staff and the group of Prof. Pettenkofer (HZB) for assistance during beamtime.

6. References

[1] Int. Technol. Roadmap for Semiconductors 2007 Edition for "Emerging Research Devices" at http://www.itrs.net/Links/2007ITRS/Home2007.htm
[2] Richard S, Aniel F, and Fishman G 2004 *Phys. Rev. B* ***70*** 235204.
[3] Belabbes A, Zaoui A, and Ferhat M 2007 *Materials Science and Engineering B* **137** 210.
[4] Hasegawa H, Akazawa M, Domanowska A, Adamowicz B 2010 *Appl. Surf. Sci.* **256** 5698.
[5] Ye PD, Wilk GD, Yang B, Kwo J, Chu SNG, Nakahara S, Gossmann HJL, Mannaerts JP, Hong M, Ng KK, and Bude J 2003 *Appl. Phys. Lett.* **83** 180.
[6] Frank M, Wilk GD, Starodub D and Gustafsson T 2005 *Appl. Phys. Lett.* **86** 152914.
[7] Hinkle CL, Sonnet AM, Vogel EM, Mcdonnell S, Hughes GJ, Milojevic M, Lee B, Aguirre-Tostado FS, Choi KJ, Kim HC, Kim J, and Wallace RM 2008 *Appl. Phys. Lett.* **92** 071911.
[8] Tallarida M, Adelmann C, Delabie A, Van Elshocht S, Caymax M and Schmeisser D 2011 *Appl. Phys. Lett.* **99** 042916.
[9] Tallarida M, Weisheit M, Kolanek K, Michling M, Engelmann HJ and Schmeisser D 2011 *J. Nanopart. Res.* **13** 5975.
[10] Tanuma S, Powell CJ, and Penn DR 1994 *Surf. Interf. Anal.* **21** 165.
[11] Coulson CA, Rèdei LB and Stocker D 1962 *Proceedings of the Royal Society of London. Series A, Mathematical and Physical Sciences* **270** 357.
[12] Klejna S, Elliott SD 2012 *J. Phys. Chem. C* **116** 643.

IOP Conf. Series: Materials Science and Engineering **41** (2012) 012004 doi:10.1088/1757-899X/41/1/012004

The effect of composition on the bandgap width in insulating $Nb_xTa_yO_z$ nanolayers

W. C. Wang[1], H. Y. Chou[1], M. Badylevich1[1, 2], T. Blomberg[3], Ch. Wenger[4], J. A. Kittl[5] and V. V. Afanas'ev[1]

1. Department of Physics and Astronomy, University of Leuven, Belgium

2. Present affiliation: Light and Lighting Laboratory, Catholic University College Gent, Belgium

3. ASM Microchemistry Ltd., 00560 Helsinki, Finland

4. IHP, 15236 Frankfurt (Oder), Germany

5. IMEC, B-3001 Leuven, Belgium

e-mail: WanChih.Wang@fys.kuleuven.be

Abstract. The bandgap width in niobium tantalate ($Nb_xTa_yO_z$) films with Ta/(Ta+Nb) ratio of 10.7, 41.9, and 68.4 % are studied in this work. The samples were prepared by atomic layer deposition on (100)Si/TaN substrates, some of them subjected to crystallization annealing. Two photoconductivity spectral thresholds are found in all the studied samples: one at around 4.2 eV and another in range 3.2-3.8 eV, which may be correlated with the bandgaps associated with TaO_x and NbO_x sub-networks, respectively. The observed preservation of the narrowest bandgap of NbO_x despite the addition of the Ta cation indicates that electron states of Nb and Ta have insufficient mutual interaction to affect the gap width of two oxide sub-networks.

1. Introduction

The aggressive down-scaling of electronic devices prompts the search for insulators with high permittivity (κ) value to reach a lower equivalent oxide thickness (EOT). By allowing one to use the physically thicker layer, the high-κ materials suppress the direct tunneling leakage currents as compared to the conventional SiO_2 layer of the same EOT. However, the leakage caused by trap-assisted-tunneling and the lower barriers for electrons due to the relatively narrow bandgap (E_g) of high-κ insulators remain problems to be solved.

So far the efforts to develop a material combining high κ value with a wide E_g merely encountered a success.[1] One of the possible approaches to address this issue consists in addition of another cation to an existing high-κ oxide in a hope to combine the best properties of two constituents. In this work, the effect of adding Ta cation to Nb_2O_5 on band structure is studied. The initial intention was to combine the wide E_g of Ta_2O_5 and the high-κ value of Nb_2O_5 in one complex oxide. We found, however, that $Nb_xTa_yO_z$ films of different Ta to Nb ratio exhibit similar E_g in both amorphous and crystallized phases, with the smallest E_g correlated with NbO_x network is still observed while for all compositions the κ value of the mixed oxide does not exceed the permittivity of Ta_2O_5.

2. Experimental

The 45 nm thick $Nb_xTa_yO_z$ layers were fabricated by using atomic layer deposition (ALD) technique in ASM Pulsar 2000 R&D reactor at 225 °C on the top of physical vapor deposited TaN bottom electrodes on p-type (100)Si substrates. TaF_5 and NbF_5 were used as metal precursors and a combination of H_2O and O_3 was used as the oxidizer. The relative Ta content Ta/(Ta+Nb) in the $Nb_xTa_yO_z$ films was 10.7, 41.9, and 68.4 %. To study the effect of crystallinzation, samples of each composition were subjected to the post-deposition annealing (PDA) at 600 °C in N_2 ambient for 5 min.[2] Then, the 15-nm thick Au electrodes of 0.5 mm area were fabricated on the $Nb_xTa_yO_z$ layers by thermal evaporation through the shadow mask to form the metal-insulator-metal (MIM) capacitor structures.

The bandgap width in the studied oxides was measured using photoconductivity (PC) spectroscopy at room temperature in the photon energy (hν) range 1.5-6.0 eV with a fixed 2 nm spectral resolution. During the PC experiments, a bias voltage (V_g) was applied on the top Au electrode of the MIM samples. The quantum yield (Y) was calculated as the photocurrent normalized to the incident photon flux[3]. The bandgap was determined as the spectral threshold of the oxide PC.

3. Results and Discussion

The E_g of the as-deposited amorphous $Nb_xTa_yO_z$ (a-$Nb_xTa_yO_z$) films with Ta/(Ta+Nb)=10.7 % was extracted from the $Y^{1/2}$-hν plots, as shown in FIG. 1(a). The PC threshold is clearly seen at hν=4.0 eV from the spectral curve measured under a low negative bias voltage (V_g=-0.1 V) applied on the top Au

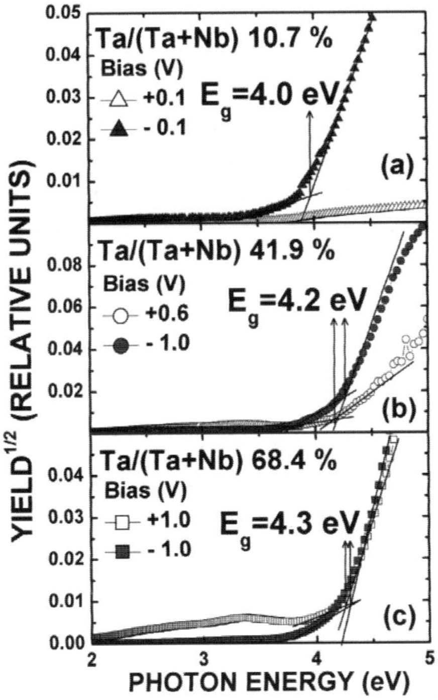

FIG. 1. The PC spectra plotted in $Y^{1/2}$-hν coordinates illustrating the determination of the bandgap of the amorphous $Nb_xTa_yO_z$ layers from the PC spectral threshold. The open and solid symbols represent the data measured under positive and negative bias, respectively.

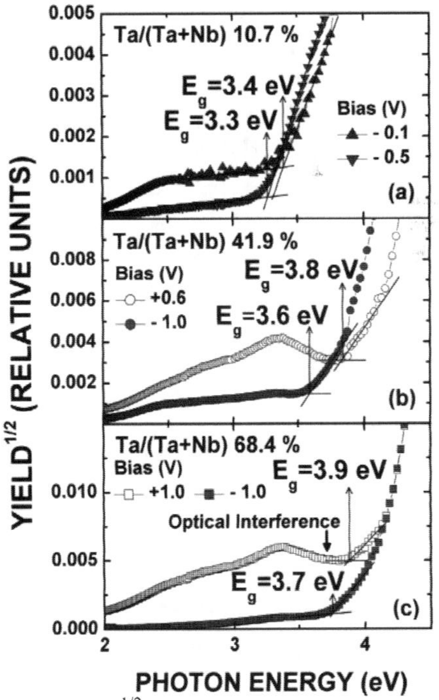

FIG. 2. The $Y^{1/2}$-hν plots of PC results illustrate the determination of the lower bandgap of the a-$Nb_xTa_yO_z$ layers from the PC spectral threshold. The open and solid symbols represent the data measured under positive and negative bias, respectively.

electrode (▲). However, when the sample is biassed positively (V_g=0.1 V), the photocurrent still flows in the same direction as it does under the negative bias (△), i.e. the generated electrons drift from Au towards TiN. This is due to the high work function (WF) of Au (5.1-5.3 eV) as compared to the WF of TaN (4.5-4.7 eV), resulting in a negative built-in voltage across the $Nb_xTa_yO_z$ layers.

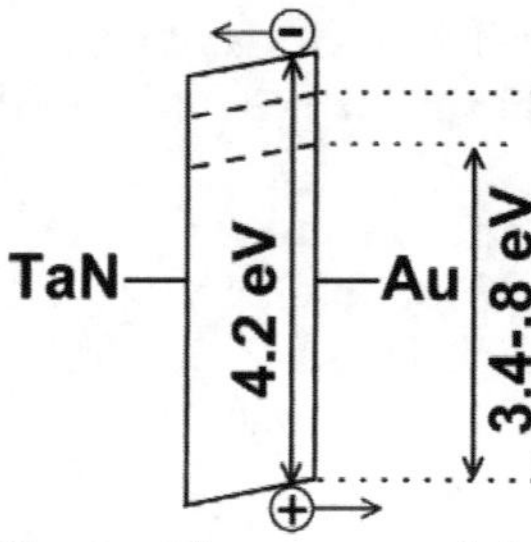

FIG. 3. The proposed band diagram of a-$Nb_xTa_yO_z$ layers under zero bias condition.

The PC spectra measured under both positive and negative bias on MIM samples with a-$Nb_xTa_yO_z$ insulators with higher Ta content (Ta/(Ta+Nb)=41.9 and 68.4 %) reveal spectral thresholds in the range 4.2-4.3 eV as illustrated in Fig. 1(b) and Fig. 1(c). Therefore, we may conclude that in the studied range of Ta content from 10.7 to 68.4 % the PC threshold of 4.2±0.1 eV is found. This threshold is close to the PC threshold in a-Ta_2O_5 (E_g=4.4 eV[3]), suggesting that, in terms of electron structure, the Ta-oxide sub-network in a-$Nb_xTa_yO_z$ layers resembles that of amorphous Ta oxide.

When zooming in the low-energy portion of the PC spectra, we also found PC thresholds in the hν range below the already discussed PC threshold of the TaO_x sub-network. As can be seen from the $Y^{1/2}$-hν plots shown in Fig. 2 for samples with different Ta content, the corresponding spectral thresholds gradually shift form E_g~3.3 eV (Ta/(Ta+Nb)=10.7 %) to higher values reaching ~3.9 eV for Ta/(Ta+Nb) equal to 68.4 %. This behavior may be explained by the presence of a phase(s) resembling crystallized Nb_2O_5 (E_g=3.4 eV[4]) in the Nb-rich sample. Though the optical interference in the insulator hampers the accurate determination of the PC threshold in the layers with a higher Ta content, particularly under positive bias, the increase of the gap width is clearly visible in Fig. 2 (b, c). This effect may be associated with the formation of NbO_x sub-network with electronic structure resembling that of a-Nb_2O_5 (E_g~4.0 eV[6]).

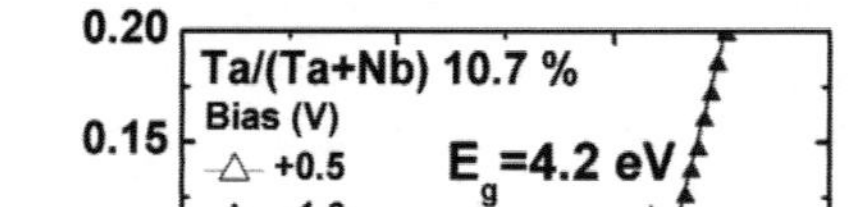

FIG. 4. The PC spectra plotted in $Y^{1/2}$-hν coordinates illustrating the determination of the bandgap of the crystallized $Nb_xTa_yO_z$ layers from the PC spectral threshold. The open and solid symbols represent the data measured under positive and negative bias, respectively.

As the Ta/(Ta+Nb) ratio becomes larger, we encountered a reduction of leakage (dark) current which is consistent with the electron conduction determined sub-network with the narrowest gap present in the insulator layer, i.e. the NbO_x sub-network. The suggested gap variation scheme in the a-$Nb_xTa_yO_z$ layers is illustrated by the band diagram for zero-bias case shown in FIG. 3 where the valence band top of the TaO_x and NbO_x sub-networks are supposed to be at the same energy[7].

The effect of annealing $Nb_xTa_yO_z$ layers on E_g is studied by analyzing the PC spectra of the crystallized $Nb_xTa_yO_z$ (c-$Nb_xTa_yO_z$) layers as illustrated in FIG. 4. Since the leakage current decreases after annealing in c-$Nb_xTa_yO_z$ films of Ta/(Ta+Nb) ratio=10.7 %, we were able to observe PC signals under opposite voltage polarities revealing a slight increase of the E_g value to around 4.2 eV.

The same 4.2-eV PC threshold is also found in the c-$Nb_xTa_yO_z$ layers with a higher Ta content as illustrated in Fig. 4(b, c), though in the MIM capacitor with Ta/(Ta+Nb)=41.9 % the PC spectrum can be measured only if a negative bias is applied to the top metal electrode. Thus, summarizing the results for both a- and c- $Nb_xTa_yO_z$ films, the spectral threshold of PC is found in all samples at 4.2±0.1 eV, apparently due to the development of electronic structure of TaO_x sub-network similar to the a-Ta_2O_5 one.

FIG. 5 shows the $Y^{1/2}$-hν plots of c-$Nb_xTa_yO_z$ layers measured in the lower photon energy range. The PC spectra measured under negative biases (filled symbols) reveal the same trend in the variation of the PC spectral theresholds as in the unannealed amorphous layers: The Nb-rich film exhibits E_g=3.2 eV, which is seen to increase to ~3.8 eV with the increasing Ta content. Interestingly, however, is that the PC spectral obtained with positive bias applied to the top Au electrode (open symbols) do not show the lowest 3.2-eV threshold. These observations might point towards the in-depth inhomogeneity of the $Nb_xTa_yO_z$ films with Nb-rich and Ta-rich layers located near the opposite electrodes of MIM capacitors. Moreover, the spectra also indicate the development of a photocurrent even at photon energy hν <3 eV. The spectral threshold of this photoexcitation is found to be close to ~2 eV and may be related to the presence of defect levels in the insulating layers.

4. Conclusion

Two energy gaps are seen both in the amorphous and crystallized $Nb_xTa_yO_z$ samples with Ta/(Ta+Nb) ratio in the range from 10.7 to 68.9 %: The wider E_g of 4.2±0.1 eV may be associated with the TaO_x sub-network, while the narrower E_g ranging from 3.2 to 3.8 eV correlates with Nb content and, therefore, is suggested to be related to NbO_x sub-network. Since the narrower E_g of NbO_x is still observed, the addition of the second Ta cation is seen to be inefficient to attain an increase of the bandgap width. Therefore, for the studied ALD oxides, the PC results indicate that mixing Nb_2O_5 with Ta_2O_5 delivers no substantial benefit in optimization of the insulating properties.

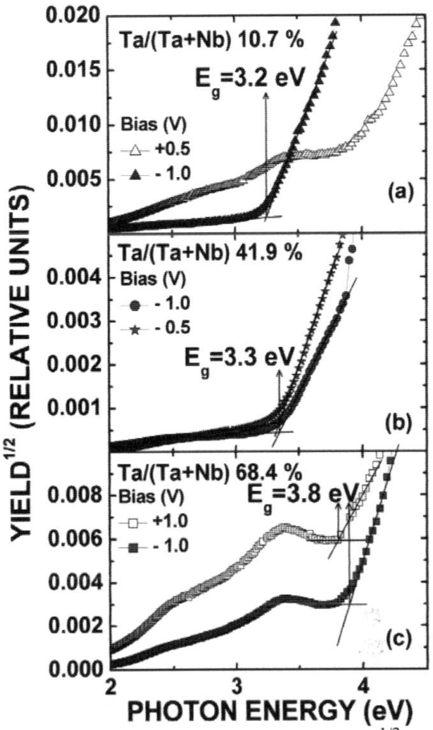

FIG. 5. The PC spectra plotted in $Y^{1/2}$-hν coordinates illustrating the determination of the lower bandgap of the c-$Nb_xTa_yO_z$ layers from the PC spectral threshold. The open and solid symbols represent the data measured under positive and negative bias, respectively.

References

[1] Kittl J A, Opsomer K, Popovici M, Menou N, Kaczer B, Wang X P, Adelmann C, Pawlak M A, Tomida K, and Rothschild A 2009 Microelectron. Eng. **86** 1789
[2] Blomberg T, Wenger Ch, Baristiran Kaynak C, Ruhl G, and Baumann P 2011 Microelectron. Eng. **88** 2447
[3] Afanas'ev V V and Stesmans A 2007 J. Appl. Phys. **102** 081301
[4] Norton D 2004 Mater. Sci. Eng. R **43** 139
[5] Kukli K, Ritala M, and Leskela M 2001 J. Electrochem. Soc. **148** F35
[6] Afanas'ev V V and Stesmans A, in High-κ gate dielectrics, edited by Houssa M (IOP, Bristol, 2004), p. 223
[7] Afanas'ev V V, Houssa M and Stesmans A 2011 Integr. Ferroelectr. **125** 53

E-MRS 2012 Spring Meeting: Symposium M IOP Publishing
IOP Conf. Series: Materials Science and Engineering **41** (2012) 012005 doi:10.1088/1757-899X/41/1/012005

Sacrificial ion beam etching process for seed layer removal of 6 µm pitch CuSn micro bumps

J Hess, H Vogt

Fraunhofer Institute for Microelectronic Circuits and Systems, Finkenstrae 61, 47057 Duisburg, Germany

E-mail: jennifer.hess@ims.fraunhofer.de

Abstract. Copper pillar bumps show a wide-ranging application for assembly and packaging according to the "More than Moore" roadmap. For the demand of higher input/output (I/O) densities and consequently smaller bump pitches the requirements on each process step in producing 6 µm pitch Cu-Sn bumps increase. In this case the removal of seed layer with wet etchants is no longer practicable due to high undercut. A sacrifical Ion Beam Etching (IBE) process was developed for removing the TiW/Cu seed layer without any undercut. Due to the high etching rate of the rough Sn surface a sacrificial layer of Ni was used to protect the solder layer. To optimize the layer thicknesses etch rates were characterized. Special attention was directed to the etched material which covered the bumps on the sidewalls after the etching process step. Energy-dispersive X-ray spectroscopy (EDX) measurements and reflow processes revealed the influence of the redepositioned material on the melting behavior and hence on the following bonding process.

1. Introduction

As part of the general trend of smaller and cheaper interconnect technologies copper pillars enter the market as promising candidates. Compared to solder bumps copper pillars show good advantages of better electrical properties and higher stand-off height for underfill processes. Smaller bump pitches and hence a higher number of IO counts can be achieved. Their application field ranges from low cost flip chip packages to 3D integration stacks. Amkor Technology already showed 40 µm pitch copper pillars in fine pitch copper pillar flip chip packages [1]. In time further decrease of the interconnection pitch size is expected due to the gap in feature size between the integrated circuit and the printed circuit board. Several publications about copper pillars with bump pitches in the range of 10 µm till 30 µm can be found [2–6]. For further geometry shrinkage the limits of fabrication and soldering processes must be considered.

In this work copper pillars with a bump pitch of 6 µm were fabricated. The main focus was on the fabrication step of seed layer etching which poses a challenge especially for bump sizes smaller than 10 µm in diameter. The seed layer is an essential layer for electroplating the bump metals and has to be removed after the deposition step. The standard method of the removal is wet etching. Due to the high undercut induced by this isotropic process wet etching is no longer practicable for small bump diameter. In this work a sacrifical ion beam etching (IBE) process for removing the TiW/Cu seed layer without any undercut has been developed.

2. Process details

Figure 1 illustrates the main fabrication steps of the 6 µm copper pillars. On top of an silicon

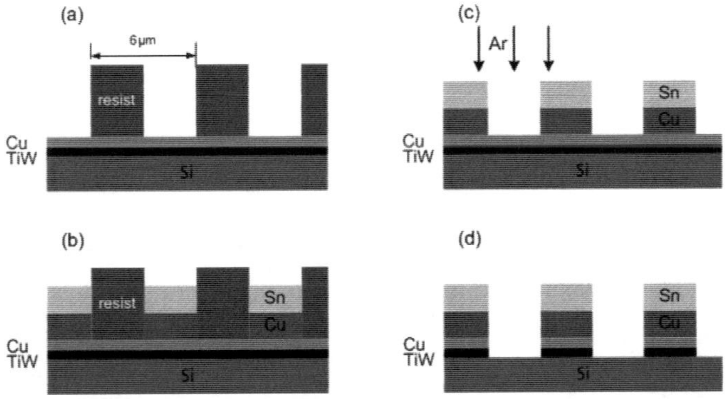

Figure 1. Main fabrication steps of 6 µm copper pillars with the solder tin. (a) Sputtering the seed layer TiW and Cu and develop a 6 µm pitch resist structure with 8 µm thickness. (b) Electroplating the bump metall Cu and Sn. (c) Remove the resist and etch the seed layer with ion beam etching and argon ions. (d) After isolating the bumps flip chip bonding can follow.

wafer the seed layers titanium tungsten (TiW) and copper (Cu) are sputtered (figure 1(a)). The thicknesses are 270 nm and 100 nm respectively. The composition of the titanium tungsten target is 10 % titanium and 90 % tungsten. 3 µm wide holes and a 6 µm pitch are generated by a photolithography step with the resist AZ10XT and the developer AZMIF826 (figure 1(b)). The resist thickness is 8 µm hence an aspect ratio of 2.7 is sucessfully developed. The next process step is the electroplating of the bump materials copper and the solder layer tin (Sn). From an sulfuric acid based smooth copper electrolyte 3 µm copper were deposited. From a matt tin electrolyte based of methanesulfonic acid 2 µm tin were achieved. After removing of the resist the important step of seed layer etching follows (figure 1(c)). Afterwards the resulting structure (figure 1(d)) can be bonded via flip chip bonding and electrically tested. In table 1 the main properties of the ion beam etching process are shown.

Table 1. Main properties of the ion beam etching experiments.

ion beam etching	
Ar flow	12 sccm
ion energy	400 eV
current density	0.4 mA/cm^2

3. Wet etching and ion beam etching

Figure 2 (a) shows a CuSn bump after the wet etching. A nearly complete undercut of the bump is visible. Only a narrow bar of the copper remains to support the upper tin layer. In the background tin layers from other bumps can be seen which are already detached from the surface. Several experiments show that the removing of the Cu/TiW seed layer is challenging. The undercut is in the range of a few micrometer (~3-5 µm) which shows that the applied etching process is no longer practicable for bump sizes in the range of 3 µm. Therfore another process to remove the metal has to be applied. Dry etching methods for the removal of copper can not be used due to the low vapor pressure compounds (CuF_x, $CuCl_x$) which copper forms with halogens. Only with elevated temperatures of about 200°C the etching is possible because above 200°C more volatile compounds can be created [7]. In comparison to the wet etching process in figure 2 (a) figure 2 (b) shows a SEM images from an ion beam etched 6 µm pitch CuSn microbump array. Instead of an undercut the seed layer remains as a base around the bump. The bumps are sufficiently isolated. Due to the physical nature of the process no undercut is detectable. Figure 3 shows the bumps before (figure 3(a)) and after (figure 3(b)) the ion beam

Figure 2. (a) Wet etching - High undercut of a CuSn microbump due to the isotropic nature of the process. (b) Ion beam etching - No undercut of a CuSn microbump array due to the physical nature of the process.

etching in more detail. The sketched arrows in figure 3(a) illustrate the direction of the argon ion beam onto the substrate and its impinge on the bumps and in between. Typical results of ion beam etched small CuSn microbumps are shown in the images 3(b) and (c). The seed layer is etched and the bumps show an incoherent tin layer with a crown formation around the top surface. As you can see on the tilted view, less material exists inside of the crown. The tin layer is almost removed and the formation of cones is visible. Cone formation occurs during the etching when material with a lower etch rate masks material with a higher rate [8]. Unfortunately the high loss of the tin material during the etching is a problem.

Figure 3. Before and after ion beam etching of CuSn microbumps: (a) before eching (b),(c) after etching.

The etching process lasts about 22 minutes based on the etching times of Cu and TiW. The etch time, seen in table 2, is determined by secondary ion mass spectrometry (SIMS). One of the corresponding SIMS-profiles is shown in figure 4. The dotted lines highlight the moment when Cu or TiW is removed.

Two problems have to be solved. 1) The high loss of tin due to its high etching rate. Due to a high sputter yield and to the roughness of the tin layer according to Cu and TiW the process can

Table 2. Etch time of 270 nm Cu and 100 nm TiW.

	etch duration
	[min]
Cu	9-10
$Ti_{10}W_{90}$	9-11

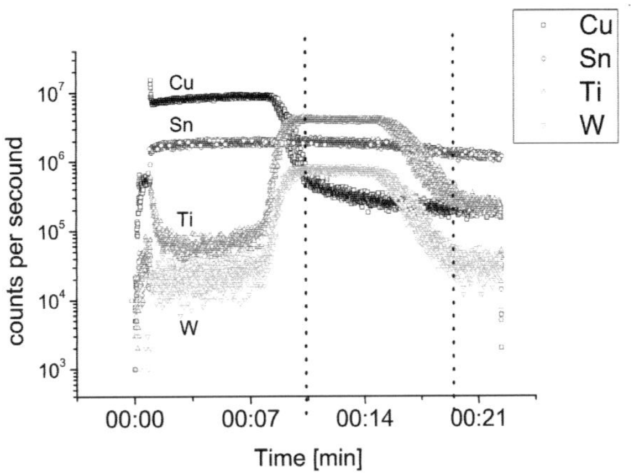

Figure 4. Recorded secondary ion mass spektrograph. The elements Cu, Sn, Ti and W are logged.

not be successful when the tin is not protected during the etching. A solution will be introduced in section 4. 2) Redepositions of sputtered material on the bump sidewalls and on the top can restrict the melting of the solder. In more detail this aspect will be reflected in section 5.

4. Sacrificial ion beam etching model

To avoid the complete loss of tin during the ion beam etching we introduce a sacrificial nickel (Ni) layer which protects the tin during the etching. In figure 5 the sacrificial ion beam etching model is explained schematically. The nickel layer is deposited on the tin layer (figure 5(a)), as explained in the process details. The nickel layer needs a minimum thickness, so it is still present after the complete removal of the seed layer, see figure 5(b). Consequently the tin will be the longest possible time protected against redepositions on the bump top. In the additional etch time the nickel will be completely removed (5(c)).

The etch rate of nickel was determined based on etched structured nickel layers. In table 3 the experimental etch rates compared to literature values are summarized. The experimental etch rates from TiW and Cu originate from the etch times in table 2. Values from the literature ([9]) are generated with an argon beam and an ion energy of 500 eV and ion current density of about 1 mA/cm^2. Compared to the copper the nickel etch rate estimated in this work seems to be higher then expected from the literature values. This can be explained by the roughness of the electroplated nickel layer which can contribute to a higher etch rate. Based on an etch time of 22 minutes we calculate a nickel thickness of 660 nm which can be removed within that time. According to our model an additional nickel thickness is necessary. In conclusion a 800 nm nickel

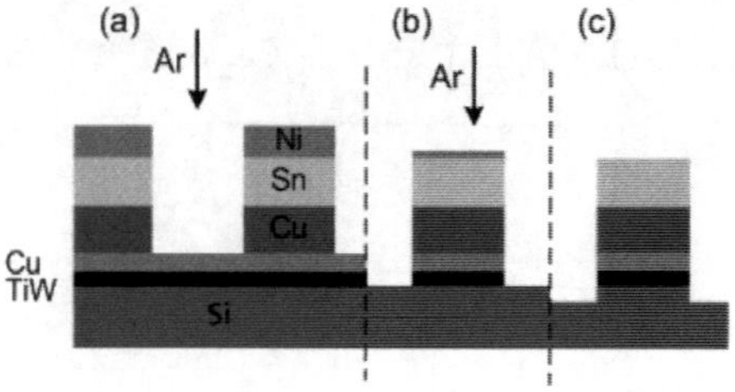

Figure 5. Sacrificial ion beam etching model. (a) The nickel protects the tin layer during the etch process. (b) After the removal of the seed layer the nickel should still be present and protect the tin. (c) In the additional etch time the nickel will be removed.

Table 3. Etch rates based on experimental data and literature. The literature values ([9]) from TiW are individually given for each element.

	etch rates	
	in this work	lit.
	400 eV, 0.4 mA/cm^2 [nm/min]	500 eV, 1 mA/cm^2 [nm/min]
Cu	~30	110
$Ti_{10}W_{90}$	~10	38
Ni	~30	66

layer is suitable for etching the CuSn microbumps. Figure 6 shows the SEM images of the CuSn microbumps with the nickel sacrificial layer before and after the etching. After the removal of the seed layer the tin is still present and the nickel layer was successfully used as protector.

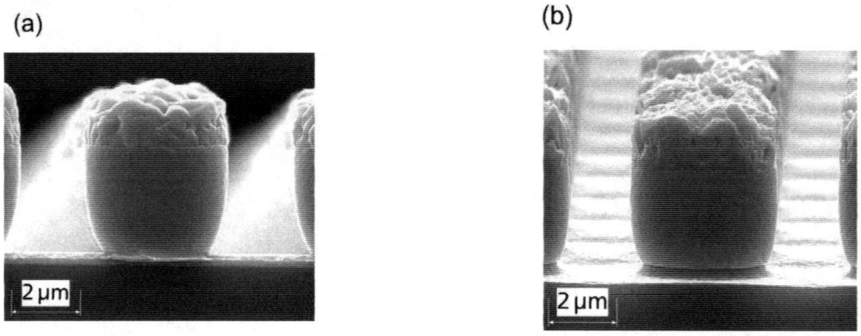

Figure 6. Applying the sacrificial ion beam etching model to 6 μm pitch CuSn microbumps. (a) before etching. (b) after etching.

5. Redepositions

The images of etched CuSn microbumps, shown in figure 3(b), illustrate the problem with redepositions. Sputtered material can deposit again on the sidewalls or the top of the bumps. All elements participating in the structure can be detected with energy-dispersive X-ray (EDX) measurements. Figure 7 shows that redeposited material can inhibit the melting of the solder layer. A part of the tin layer melts and gather in the center of the bump. Insoluble material based on the redepositions remain at the top of the bump. To avoid redepositions several

Figure 7. CuSn microbumps after ion beam etching and heating up to 350°C. Insoluble material remains at the top whereas liquid tin gathers in the bump center.

approaches are presented. (1) A decrease of the seed layer thickness will efficiently reduce the amount of redepositions. A decreasing of more than 50% is being considered. (2) The reactive ion beam etching (RIBE) or the chemically assisted ion beam etching (CAIBE) of TiW using the gases SF_6 or CF_4 are promising. With a more chemical assisted process it is likely possible to fully remove the sputtered material and avoid the redeposition. (3) With changing the angle of incidence (Θ) to higher angles the removing of the redopositions during the eching can be tested (see the schematic model in figure 8). A higher angle to the top surface causes a higher etch rate and an angle less than 90° at the sidewalls allows the etching and therefore the removing of the redepositions. Due to shadowing the seed layer by the bumps a two step etch process can be considered. First of all the seed layer can be significantly reduced by etching with normal incidence then the redoposition can be removed by a higher angle of incidence.

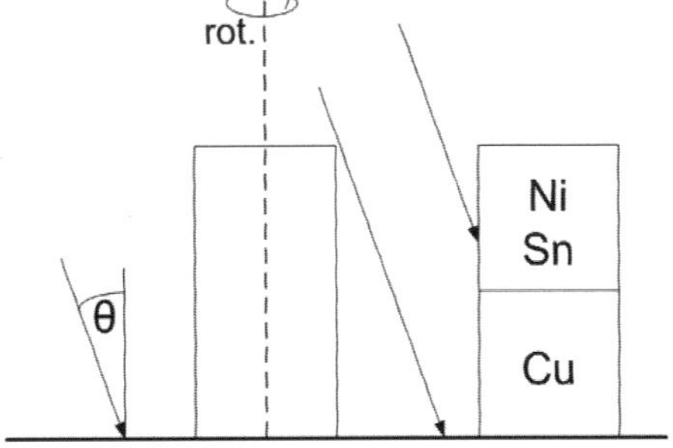

Figure 8. Introduced model for etching the structure with an higher angle of incidence Θ.

6. Conclusion

The fabrication of 6 µm pitch Cu-Sn bumps was introduced. The removing of the Cu/TiW- seed layer was discussed in detail. As part of the miniaturization of the copper pillars the removing the seed layer with wet etching methods is challenging therefore a ion beam etching method was used which allows etching without undercut. A sacrificial Ion Beam Etching (IBE) process was developed for removing the seed layer without any undercut and the protection of the tin layer.

A nickel layer was used to protect the rough tin layer for removal. The nickel thickness should be in the range of 800 nm to utilize as perfect protection. Redepositions on the sidewalls and on the top of the bumps inhibit the melting of the solder. They should be avoided in the future by applying one of the three different methods which were introduced to minimize the re-deposited material. After the successful development of the etch process flip chip bonding and electrical tests will be performed.

References

[1] "Amkor Technology and Texas Instruments Deliver Industrys First Fine Pitch Copper Pillar Flip Chip Packages to Market", Press Release, July 7, 2010

[2] Reed J, Lueck M, Gregory C, Huffman A, Lannon Jr J and Temple D 2010 Low temperature bonding of high density large area array interconnects for 3d integration *43rd IMAPS Conference, Raleigh, NC November* vol 2

[3] Yu A, Lau J, Ho S, Kumar A, Hnin W, Yu D, Jong M, Kripesh V, Pinjala D and Kwong D 2009 Study of 15 µm pitch solder microbumps for 3d ic integration *Electronic Components and Technology Conference, 2009. ECTC 2009. 59th* (IEEE) pp 6–10

[4] Tanida K, Umemoto M, Tanaka N, Tomita Y and Takahashi K 2004 *Japanese journal of applied physics* **43** 2264

[5] Zhan C, Chuang C, Juang J, Lu S and Chang T 2010 Assembly and reliability characterization of 3d chip stacking with 30μm pitch lead-free solder micro bump interconnection *Electronic Components and Technology Conference (ECTC), 2010 Proceedings 60th* (IEEE) pp 1043–1049

[6] Huebner H, Penka S, Barchmann B, Eigner M, Gruber W, Nobis M, Janka S, Kristen G and Schneegans M 2006 *Microelectronic engineering* **83** 2155–2162

[7] Steinbrüchel C and Chin B 2001 *Copper interconnect technology* vol 46 (Society of Photo Optical)

[8] Stewart A and Thompson M 1969 *Journal of Materials Science* **4** 56–60

[9] Williams K and Muller R 1996 *Microelectromechanical Systems, Journal of* **5** 256–269

IOP Conf. Series: Materials Science and Engineering **41** (2012) 012006 doi:10.1088/1757-899X/41/1/012006

Atomic Layer Deposition of SiN for spacer applications in high-end logic devices

F Koehler, D H Triyoso, I Hussain, S Mutas and H Bernhardt

GLOBALFOUNDRIES, Wilschdorfer Landstr. 101, 01109 Dresden, Germany

E-mail: fabian.koehler@globalfoundries.com

Abstract. Continuous down scaling of transistor size resulted in introduction of high-k material to replace the conventional gate oxide. High-k Metal Gate implementation brings new challenges to the subsequent processing steps of the device. Both oxide ingress and removal of the HKMG stack during later cleaning steps must strictly be avoided. We have developed and optimized an ALD SiN encapsulation liner which protects the HKMG stack and also serves as a robust spacer for subsequent ion implantation. SiN was chosen because of its oxide-blocking characteristics and its durability in diluted hydrofluoric acid. We developed sophisticated ALD processes using ionized ammonia radicals and dichloro-silane. This ALD process has several advantages compared to conventional LPCVD or PECVD processes: Superior thickness control and uniformity, excellent step coverage and no loading effects. In addition, process temperature could be reduced to 500°C, which is shown to be critical for protecting the HKMG stack. Further reduction in process temperature to 400°C is also demonstrated using a novel precursor.

1. Introduction

Continuously increasing the number of transistors per chip is essential for the success of semiconductor industry. The subsequent reduction of transistor size has come to a critical point as the conventional gate oxide has reached its scaling limit. The nitrided silicon oxide can be scaled down to a value of about 1nm, but at further down scaling direct tunneling increases drastically resulting in increased leakage current and power consumption [1].

To fulfill the requirements of further performance increase GLOBALFOUNDRIES implements high-*k* metal gate (HKMG) in 32/28nm technology node [2]. As illustrated schematically in Figure 1 HfO replaces the conventional SiON gate oxide. The thickness of HfO layer can be increased to values larger than d = 2nm, which reduces unintentional tunneling effectively, while the capacitance of the device stays sufficiently high due to the larger k value. In the first approximation, the gate stack can be regarded as a parallel plate capacitor with its capacitance

$$C = k\, k_0\, A / d \qquad (1)$$

where k is the relative permittivity of the gate material, k_0 is the permittivity of the vacuum, A is the area of the device and d is the thickness of the gate material.

As can be seen from (1) implementation of high-*k* allows both further shrinkage of transistor size and increase of gate dielectric thickness. The latter reduces gate leakage and therefore power consumption. This is very critical for the fast-growing market of mobile products like notebooks and smartphones with extended battery life. GLOBALFOUNDRIES uses a "Gate First" approach. The "Gate First" allows further transistor scaling, minimizing die size and increasing power efficiency.

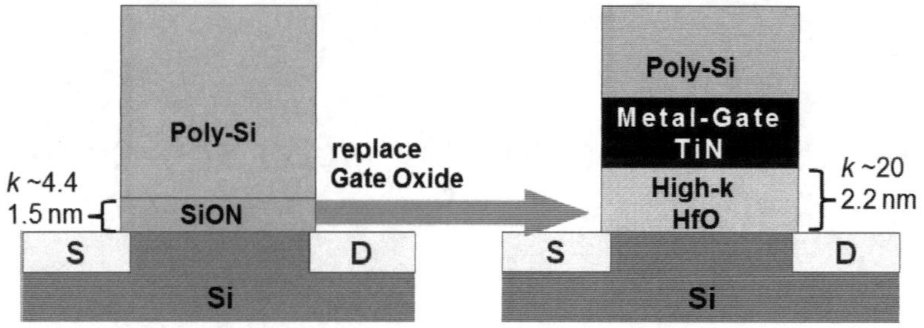

Figure 1. Replacement of the gate oxide by a HKMG stack, schematically.

2. Materials and processes for spacer applications.

Spacers are typically SiN or SiO_2 fabricated by LPCVD or PECVD [3]. Standard LPCVD SiN is deposited at relatively high temperatures (750°C) and thus often not applicable due to thermal budget restrictions. It is replaced by PECVD, which can be deposited even at moderate temperatures (400°C), but suffers from strong loading effects. ALD has the advantage of good conformality at low temperatures [4, 5]. Moreover, ALD shows superior wafer-to-wafer and within-wafer uniformity: For a SiN layer with a thickness of $d < 5$nm our data show that all values for wafer mean are within d +/- 2.5% x d. The standard deviation for wafer mean is only 1.4% x d. The within wafer standard deviation (13 points across the wafer were measured by spectroscopic ellipsometry) is 0.8% x d. For a layer of approx. 20nm we find that all wafer mean values are within d +/- 2.0% x d, standard deviation for wafer mean is 1.1% x d and within wafer standard deviation is 0.7% x d.

3. Encapsulation of the high-k metal-gate stack

TiN and HfO are critical materials concerning oxygen ingress and removal during later cleaning steps. Therefore it is necessary to protect these layers efficiently. We have developed a SiN ALD process for the gate encapsulation, which is applied directly after gate etch, as illustrated in Figure 2. A cross flow batch reactor (see Figure 3) is used for processing up to 100 wafers at the same time. The process is targeted to a thickness of < 5nm and consists of alternating pulses of NH_3 and dichloro-silane (DCS). Process temperature is 500°C, which is significantly lower compared to standard LPCVD Nitride (750°C). NH_3 plasma is activated by an RF source outside the reaction chamber (see Figure 3).

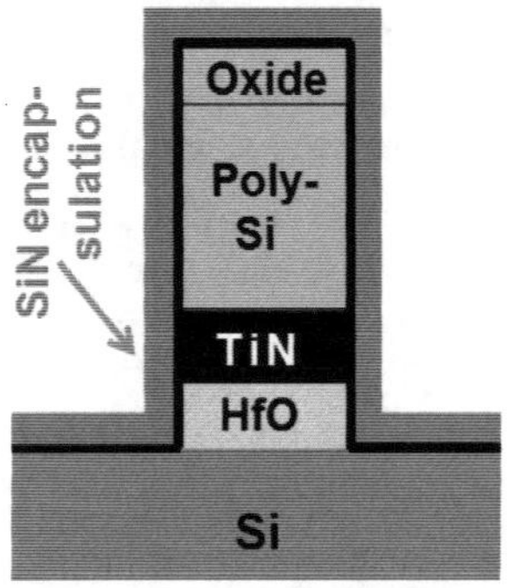

Figure 2. Gate encapsulation.

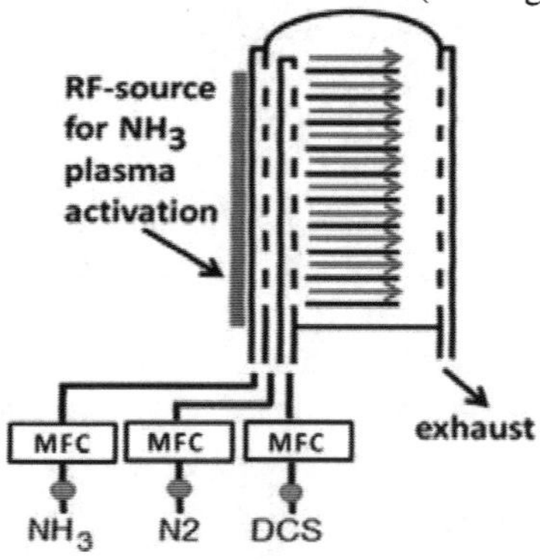

Figure 3. Cross-flow reactor.

4. Spacer formation

After deposition of an LPCVD TEOS furnace oxide liner the spacer zero etch follows, as illustrated in Figure 4. The spacer formation is realized by a plasma-enhanced dry etch process. The advantage of

IOP Conf. Series: Materials Science and Engineering **41** (2012) 012006 doi:10.1088/1757-899X/41/1/012006

this process is its strong anisotropic behavior. This spacer determines the placement of both extension- and halo-implants. A cross section TEM of spacer zero is shown in Figure 5. The spacer one is deposited at 400°C using NH_3 and a novel low-temperature precursor, monochloro-silane. The process is targeted to achieve a thickness of approximately 20nm. After spacer one etch the source/drain implants follow, as illustrated in Figure 6. Figure 7 shows a cross section TEM of Spacer 1.

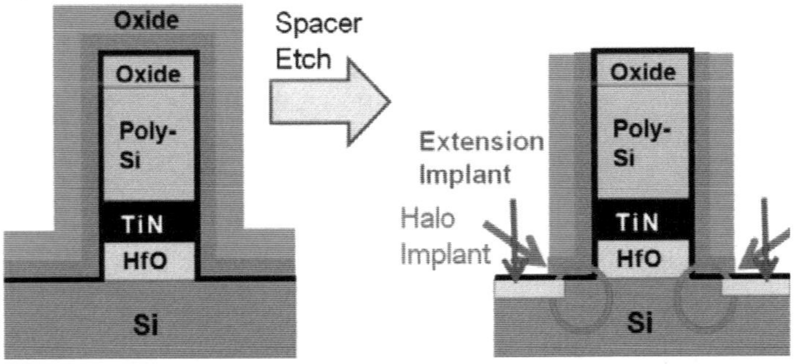

Figure 4. Spacer zero formation.

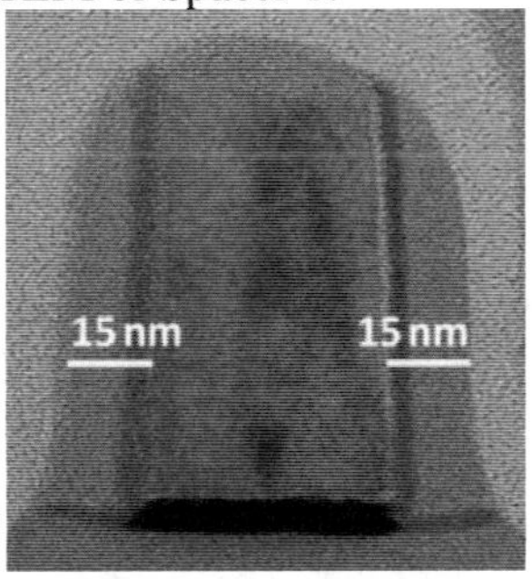

Figure 5. TEM view.

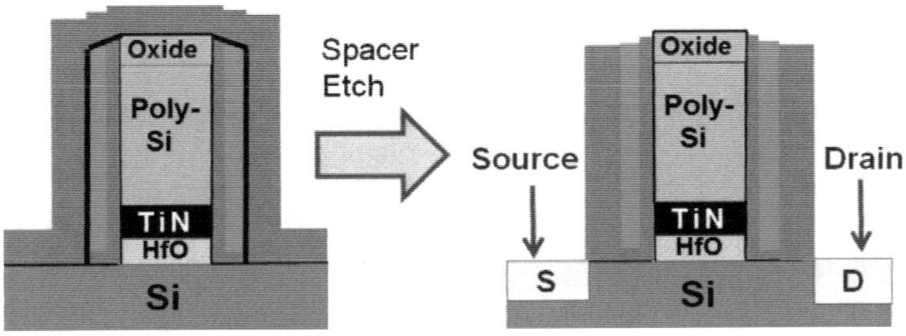

Figure 6. Spacer one formation.

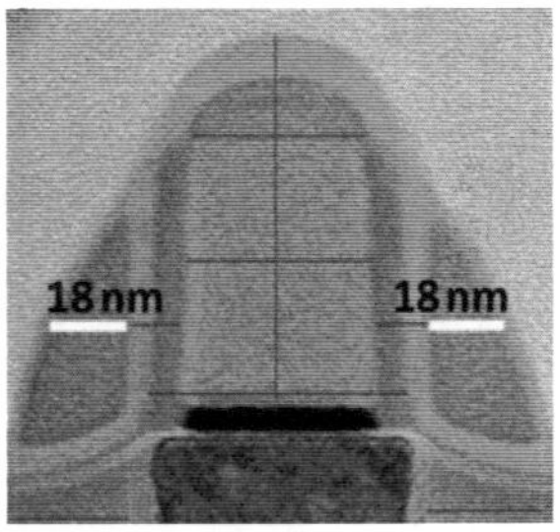

Figure 7. TEM view.

The vital advantage of ALD SiN compared to PECVD can be seen in Figures 8: The spacer thickness is independent of the transistor density. This is very important as design rules allow regions with different densities of transistors on the chip, while electrical parameters must match for all transistors.

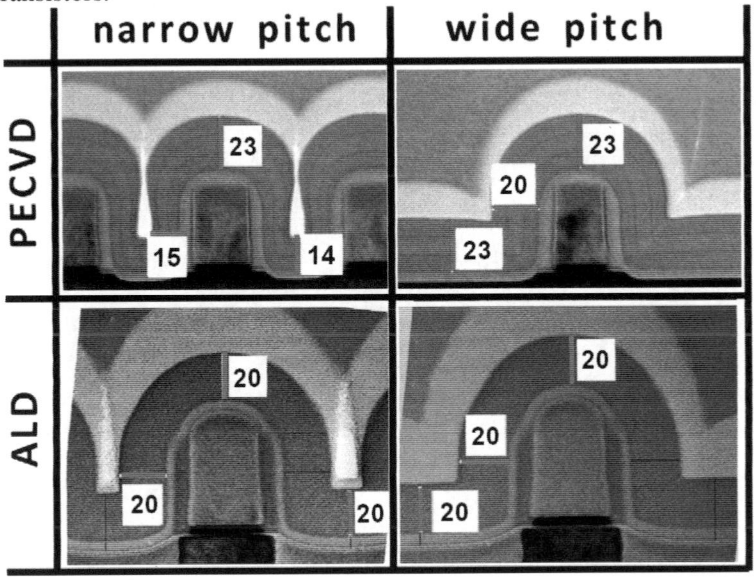

Figure 8. Comparison of PECVD and ALD spacers after deposition for narrow pitch and wide pitch regions. Strong loading effects were observed for PECVD, whereas for ALD the layer is uniform independent of transistor density. All scaling is in nm.

IOP Conf. Series: Materials Science and Engineering **41** (2012) 012006 doi:10.1088/1757-899X/41/1/012006

5. Wet etch rates

To evaluate the wet etch rates we have dipped blanket ALD SiN wafers for one minute in diluted hydrofluoric acid ($HF:H_2O$ = 1:100). Results are shown in Figure 9. As can bee seen, for temperatures below 500°C it is favorable to use the low-temperature precursor monochloro-silane instead of dichloro-silane to improve resistivity of the spacer material in subsequent cleaning steps.

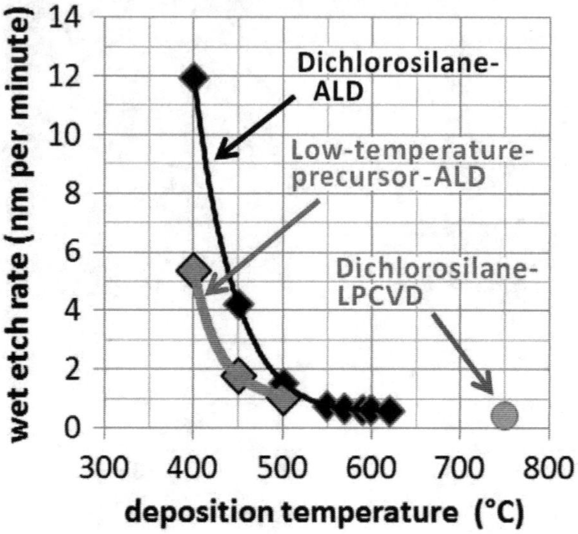

Figure 9. Wet etch rate of SiN ALD in HF : H2O = 1 : 100 as a function of deposition temperature. For comparison, LPCVD (750°C) is added.

6. Summary

We have developed ALD SiN processes at low temperatures with superior uniformity and conformality. These processes are applied in gate encapsulation and spacer formation for advanced high-*k* metal gate technology in 32/28nm node.

References

[1] M. Houssa, L. Pantisano, L.-A. Ragnarsson, R. Degraeve, T. Schram, G. Pourtois, S. De Gendt, G. Groeseneken, M.M. Heyns, 2006 "Electrical properties of high-k gate dielectrics: Challenges, current issues, and possible solutions", Materials Science and Engineering R 51 () 37–85.

[2] F. Graetsch, "New Material Challenges in 32nm Gate First High K Module", as presented in *Semicon Europa* 2011.

[3] H. Sayama, Y. Nishida, H. Oda, J. Tsuchimoto, H. Umeda, A. Teramoto, K. Eikyu, Y. Inoue and M. Inuishi, 2000, "80nm CMOSFET Technology Using Double Offset-Implanted Source/Drain Extension and Low Temperature SiN Process", IEDM Tech. Digest, pp. 239-242.

[4] M. Ritala and M. Leskela, 2001, "Atomic Layer Deposition" in Handbook of Thin Films Materials, edited by H. S. Nalwa, pp. 103-159.

[5] H. B. Profijt, S. E. Potts, M. C. M. van de Sanden, and W. M. M. Kessels, 2011, "Plasma-Assisted Atomic Layer Deposition: Basics, Opportunities, and Challenges" J. of Vac. and Science Tech. A., pp. 050801-1-050801-126.

IOP Conf. Series: Materials Science and Engineering **41** (2012) 012007 doi:10.1088/1757-899X/41/1/012007

Oxidation of Germanium and Silicon surfaces (100): a comparative study through DFT methodology

C. Mastail[1,2], I. Bourennane[1,2], A. Estève[1,2], G.Landa[1,2], M. Djafari Rouhani[1,3], N. Richard[4] and A. Hémeryck[1,2,5]

[1]CNRS, LAAS, 7 avenue du Colonel Roche, F-31400 Toulouse, France
[2]Univ de Toulouse, LAAS, F-31400 Toulouse, France
[3]Univ de Toulouse, UPS, LAAS, F-31400 Toulouse, France
[4]CEA, DAM, DIF, F-91680 Arpajon, FRANCE

[5] E-mail: anne.hemeryck@laas.fr

Abstract. Density Functional Theory calculations are used to map out the preferential oxygen molecule adsorption sites and oxygen atom incorporation on germanium (100) surface. A comparison with primary oxidation mechanisms encountered in pure silicon and silicon germanium (100) surfaces is presented here. This study highlights opposite substrates behaviors facing oxygen molecule adsorption: 1/ surface germanium atoms move from their crystalline positions to adapt to the approaching oxygen molecule resulting in adsorbed peroxide bridge configuration, whereas oxygen molecule is fully dissociated in strand configuration on a silicon surface 2/ oxygen atoms tend to avoid each other on germanium surface whereas oxide nucleus can be observed on silicon surface even at the early steps of the oxidation process. Results show that germanium surface appears to be less reactive than the silicon substrate towards molecular oxygen species.

1. Introduction

With the constant downscaling in microelectronic industry, Si-based devices reach their technological and physical limitations processes such as thermal oxidation require an ineluctable mastering of the Si/SiO_2 to pursue the silicon technological roadmap. In this context, germanium material has received a great attention as a promising candidate for replacing silicon in the next generation of complementary metal-oxide semiconductors (CMOS) technology and metal oxide field effect transistors (MOSFET). This renewed interest is motivated by the similar atomic structures and electronic properties with silicon, i.e. a small electronic band gap of 0.66 eV (1.1 eV for Si), and higher careers (3,900 cm^2/Vs for electrons and 1,900 cm^2/Vs for holes) mobility compare to silicon [1].

Moreover this material has a lattice mismatch of 4.7 % and is easily compatible with MOS Si-based technologies, and could provide emerging Si/SiGe/Ge heterostructures. However, for the required oxide layer thickness, an atomic scale understanding of the growth of the germanium oxide layer and on interfaces formation obtained during germanium oxidation process is still lacking.

The objective of this study is to establish an energetic mapping of the oxygen molecule chemical reaction as a function of its position and orientation over the germanium surface. This study has been performed using Density Functional Theory methodology. The result will be the determination of the most reactive sites and the most probable configurations that are obtained during the initial stage of the

IOP Conf. Series: Materials Science and Engineering **41** (2012) 012007 doi:10.1088/1757-899X/41/1/012007

germanium oxidation process. The relative stability of configurations containing one and two oxygen atoms incorporated into the surface bonds are characterized and commented. Additionally, we propose a comparison with the first oxygen molecule adsorption on silicon substrate during initial oxidation process. This comparative study between germanium, silicon germanium and silicon materials highlights two different behavior of both substrates and oxygen species facing the oxidation process: - contrary to the case of germanium the oxygen atoms preferentially agglomerate in the silicon oxidation case; - oxygen molecules accommodate to the silicon crystalline network whereas germanium atoms accommodate to the approaching oxygen molecule and move out from their crystalline positions.

2. Details of calculations

Substrates constructions: All surfaces considered in this article, i.e. germanium - silicon - silicon germanium surfaces, are built according to the following model. Slabs consist in periodic cells built as follow:

The germanium surface modeled with eight layers of eight germanium atoms. The cell dimensions for the germanium slab a = 16.7 Å, b = 8.2 Å, c = 21.3 Å.

For the silicon and silicon germanium surfaces, six layers of eight silicon atoms are used in the periodic cell and are detailed in the referred papers [2]. The SiGe model is built by replacing a surface silicon atom of one of the dimers by a germanium atom [3]. We checked that the difference in the chosen number of layers between Ge substrate and, Si and SiGe slabs has no impact on surface reactions and energetics. A deeper substrate was used in the case of germanium material justified by a study to come on atomic oxygen diffusion in the Ge sub-surface.

For each slab, sixteen hydrogen atoms were used in order to saturate the dangling bonds of atoms underneath the slab. The two lowest slab layers and hydrogen atoms are kept fixed during relaxation in order to simulate the bulk. All other layers, as well as the oxygen atoms, are allowed to relax. A vacuum zone of 10 Å is used in order to create a surface effect. All three surfaces have a (2x1) surface reconstruction pattern exhibiting four dimer units, two channels and a buckling along the dimer rows (Figure 1).

Resulting periodic cells are thus $Ge_{64}H_{16}$, $Si_{48}H_{16}$ and $Si_{47}GeH_{16}$ respectively for pure germanium, pure silicon and silicon/germanium slabs.

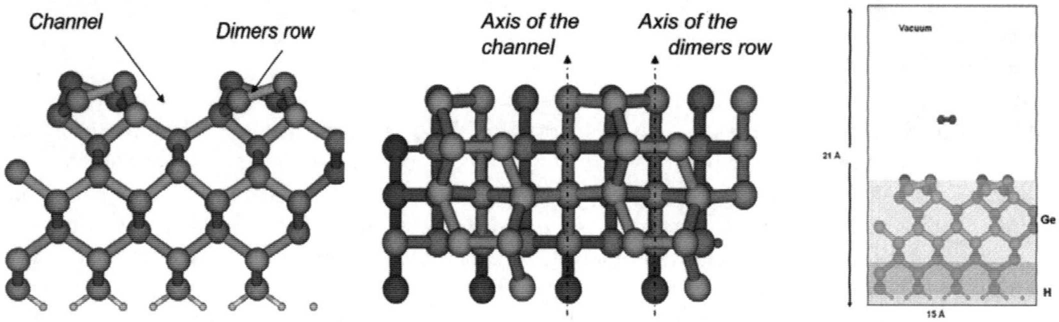

Figure 1. Side (left) and top (in the middle) views of the slab used to mimic Germanium substrate. Ge atoms are in grey and hydrogen atoms are in white. (2x1) surface reconstruction highlight dimers rows and channels. A schematic view of the periodic cell used in DFT calculations is given on the right.

Calculations Parameters. The calculations are performed in the framework of the Density Functional Theory (DFT) and implemented in the periodic VASP code [4,5]. Ultra-soft pseudo-potentials have been used with orbital developed on a basis of plane waves. For the exchange-correlation energy,

Generalized Gradient Approximation (GGA) was used and the cut-off energy fixed at 475 eV. Brillouin-zone integrations are carried out using the Monkhorst–Pack scheme [6] with a spaced mesh of 1x2x1 points in the reciprocal unit cell shifted from the origin at Γ for modeled the germanium surface. We also take into account the spin polarization to describe the oxygen molecule behavior during the adsorption process.

3. Adsorbed configurations through dry thermal oxidation

The adsorption procedure is performed by placing the oxygen molecule above the germanium surface sites with different positions and orientations initially at a height of 4 Å. At this distance, oxygen molecule and slab have no interactions. All starting configurations are schematized in Table 1. Once relaxed, the gravity centre of oxygen molecule is gradually lowered by step of 0.5 Å towards the surface until the adsorption reaction occurs. In all starting configurations, the bond axis of oxygen molecule is taken parallel to the substrate surface to favor the interaction between the occupied molecular orbital 2pΠu and the half occupied molecular orbital 2p*Πg.

Table I. Structures and energetics involved in the adsorption process on Ge(100)-(2x1). Starting and final structures are schematized. Adsorption energy gains are given in eV. White spheres represent oxygen atoms, grey squares represent germanium atoms. Grey squares size and darkness vary as a function of the layer depth from a top point of view: Larger the square is the topmost the layer.

	1	2	3	4	5	6	7	8
Starting positions	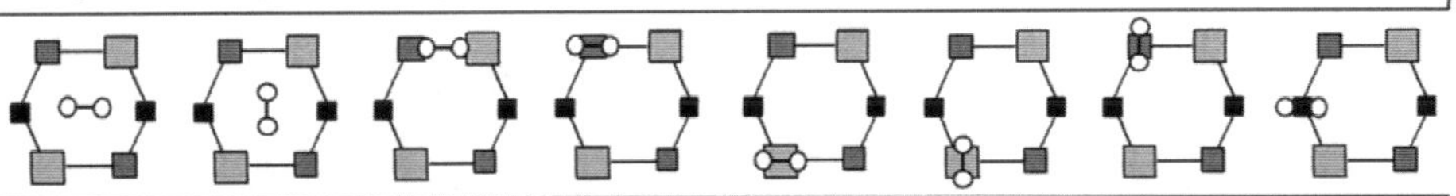							
Final configurations	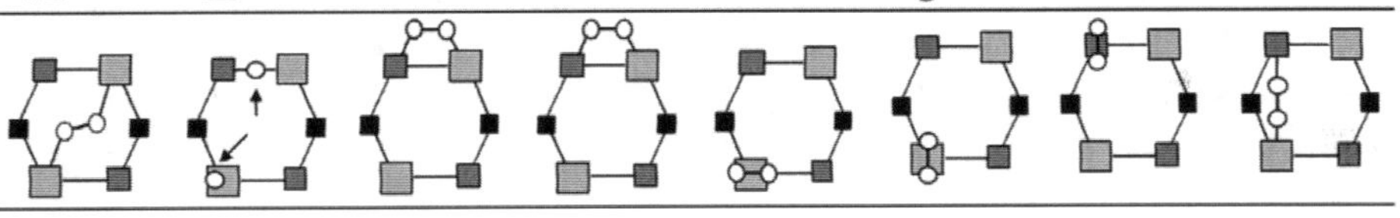							
Eads (eV)	2.00	4.81	2.79	2.79	1.64	1.64	1.27	2.54
Remarks d(O-O) (Å)	1.47	/	1.49	1.49	1.57	1.56	1.48	1.46

	9	10	11	12	13	14	15	16
Starting positions	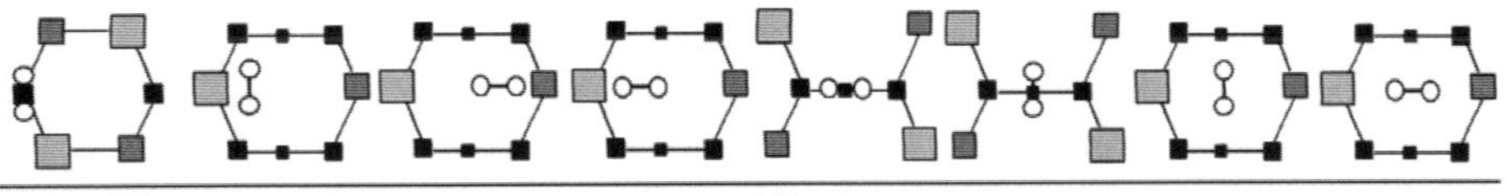							
Final configurations	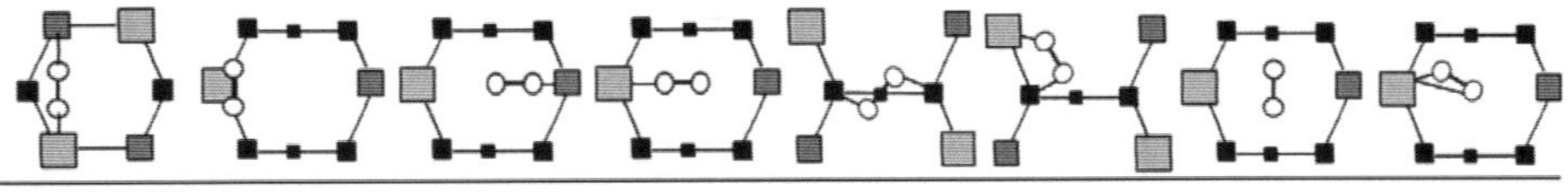							
Eads (eV)	2.56	1.63	1.43	1.32	1.54	2.31	0.39	1.63
Remarks d(O-O) (Å)	1.47	1.57	1.36	1.36	1.47	1.48	unreacted	1.56

The sixteen initial oxygen molecule positions over the germanium surface and the final adsorbed configurations are presented in Table I. Adsorption energy gains and oxygen atoms distances obtained in the final configuration are also given. Among all tested positions, only one leads to a spontaneous dissociation of the oxygen molecule. It corresponds to a narrow channel for the dissociation of one oxygen molecule initially perfectly centered between two dimers and perpendicular to the dimer row (configuration 2) [7,8]. In this adsorbed configuration, one oxygen atom is inserted into one Ge-Ge dimer bond, the second oxygen atom is stabilized in 'on top' configuration on the up-tilted atom of the adjacent dimer like a strand position as defined precisely in Ref. 9 and 10. The associated energetic gain is as large as 4.8 eV that is the highest energetic value encountered during adsorption procedure associated to the full dissociation of the oxygen molecule.

For all other adsorbed molecule configurations, a non-dissociative adsorption of the oxygen molecule is observed on the germanium surface in a good agreement with Fan's results [11]. Two configurations exhibit an adsorption of the oxygen molecule having an energetic gain of 1.3-1.4 eV (configurations 11, 12). This molecular chemisorption is characterized by a slightly stretched bond (d(O-O) = 1.36 Å) and an oxygen molecule attached to a surface germanium dimer atom by a single oxygen atom. Four configurations exhibit a partially dissociative chemisorption of the oxygen molecule on the same up-tilted germanium atom of the dimer (configurations 5, 6, 10, 16) associated with energetic gains slightly higher in the order of 1.6 eV. Both oxygen atoms are stabilized in an 'on top' configuration and are distant by 1.64 Å. We can note that this adsorption type is possible on the low-tilted germanium atom of the dimer (configuration 7). In this case, adsorption is less favorable considering its energy gain (1.27 eV) and the oxygen molecule is less stretched (d O-O = 1.48 Å). Configurations 11 and 12 will result in partially dissociative adsorbed configurations after an atomic migration of strand oxygen atom as observed for cases 5, 6, 10, 16.

Other configurations (#1, 3, 4, 8, 9, 13, 14 in Table I) exhibit a partial dissociative chemisorptions of the oxygen molecule on two distinct germanium atoms: in all these configurations, stretched or not peroxy surface bridges (Ge-O-O-Ge where d(O-O) = 1.39 to 1.5 Å) are created. These adsorbed states are the more energetically favorable non-dissociative configurations involved during adsorption. The associated energetic gain range from 2 eV to 2.8 eV depending on the adsorption occurring a single Ge-Ge dimer or on two adjacent Ge-Ge dimers: a peroxy bridge can be obtained through an adsorption of two oxygen atoms on a single dimer with an energetic gain of 2.8 eV (Configurations 3, 4). This peroxy-adsorbed state can also arise from one up and one low-tilted germanium atoms of two adjacent dimers to form an inter-dimer peroxy bridge (configuration 8, 9). An unexpected configuration is also produced by the adsorption of non-dissociated oxygen molecule on two up-tilted germanium atoms from adjacent dimers, i.e. on opposite side of the axis of the dimers row (see configuration 1). These three last cited configurations are possible thanks to a lattice deformation of the substrate. Germanium atoms bonded to the adsorbed oxygen atom move out from their crystalline site to adapt to the presence of molecular oxygen species. O-O bond is quite stretched but not dissociated with a typical length of 1.46 Å – 1.47 Å (to be compared with referred data of 1.22 Å for gaseous specie). This peroxy inter-dimer adsorbed configuration allows the linking of two germanium atoms initially distant by 4.20 Å (Configurations 8, 9) and even by 4.94 Å (configuration 1). But lattice deformation costs in energy: the more the germanium network is deformed, the lower the adsorption energy (2 eV for configuration vs 2.5 eV for configurations 8 and 9). The two positions above the channel (configurations 13, 14) can be associated to the peroxy-type adsorption except that it involves germanium atoms of the backbond of the sub-surface.

4. Discussion

Table II and III summarize all adsorption results for germanium and for silicon substrates respectively. The corresponding average energy gains and characteristic final adsorbed configurations are detailed and discussed.

We briefly detail adsorption results that we obtained in a previous study on silicon substrate [7,9]: fourteen structures exhibit an 'on top' configuration, and only one is spontaneously dissociated and incorporated within two adjacent dimers. These results are given in Table III: partially dissociative chemisorbed states have been observed with eight adsorbed states on a single dimer silicon atom, three occur upon a single surface dimer unit and three others react within two adjacent dimer units (one reacts in the channel separating the dimer rows). In the case of reacting oxygen molecule with two adjacent dimer units, a spontaneous dissociation of oxygen molecule is produced to favor an 'on top' position and keep unchanged the silicon lattice.

Table II. Summary of the oxidation of Ge(100) surface obtained in Table I. Starting positions are referred compared to starting positions described in Table I and final typical structures are schematized. Adsorption energies are given in eV. White spheres represent oxygen atoms, grey squares represent germanium atoms. Grey squares size and darkness vary as a function of the layer depth from the top: Larger the sphere is, topmost the layer.

Adsorption mechanism	Above the dimers row	Above or near a single dimer germanium atom Above the channel				
Starting Positions	2	1,13,14	8,9	3,4	5,6,7,10,16	11,12
Final Structure-types	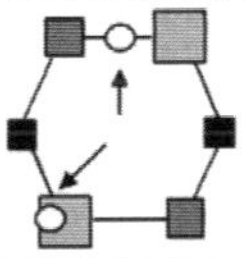	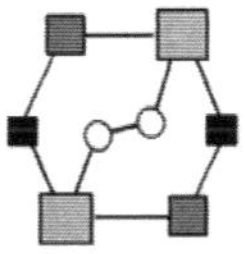	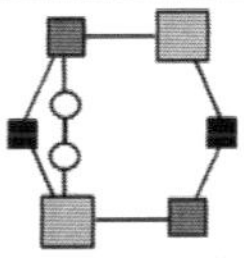	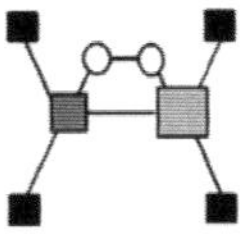	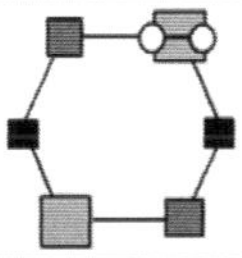	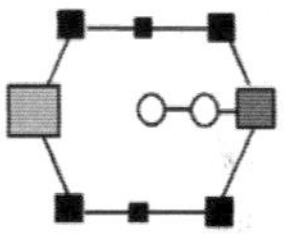
Eads (eV)	4.8	1.54 to 2.31	2.56	2.79	1.27 to 1.64	1.32 to 1.43
d(O-O) (Å)	Fully dissociated	1.47	1.47	1.49	1.48 to 1.57	1.36

At a first glance, one can notice that germanium oxidation is less exothermic than silicon oxidation process by looking at equivalent adsorption energies of comparable adsorbed structures. Both in silicon and germanium oxidation processes, the insertion of oxygen atoms into the surface bonds is energetically most favorable. Only one final configuration for germanium material leads to the full dissociation coupled to incorporation of O atom in Ge-Ge dimer bond (4.8 eV Configuration 2 Table II). This is to be compared with 6.67 eV in silicon substrate (Configuration 2 Table III) [7,9] and defined as a 'narrow barrierless channel for dissociation'.

Actually, many metastable steps are observed at the initial stages of oxidation. Table III highlights that a primary partial dissociative chemisorptions occur initially where both peroxy O-O bridge and 'on top' configurations appear to be key intermediates during germanium oxidation (Table II). On top-type of configuration was majority during silicon oxidation, whereas in the case of germanium surface, peroxy adsorption configurations are pivoted to further oxidation (further atomic diffusions deeper in the substrate, from Configuration 2 for instance). Indeed, in silicon, peroxy bridge state was only obtained when oxygen molecule was initially placed in a perfect symmetrical site above the dimer unit and only within the same dimer unit with the oxygen molecule axis parallel to the dimer axis. No adsorbed configuration in so-called inter-dimer peroxy bridge was found between Si-Si adjacent dimer bond: when a peroxy-bridge structure is observed in Germanium material, a full dissociation occurs in silicon as illustrated in Figures 2 (Four cases 1,9,13,14).

This peroxy-bridge configuration that was few observed during silicon oxidation (three cases 3,4,5), is possible thanks to the 'softness' of the germanium surface. The germanium softness allows atoms to accommodate to the electronic charge redistribution by moving out from their crystalline positions, driven by the approaching oxygen molecule interaction, giving rise to metastable adsorbed species such as inter-dimer bridges (see Figure 2). This softness is visible in all peroxy bridge configurations: for peroxy bridge configurations on a single dimer unit (# in Table II) germanium atoms are displaced to shorten the molecular oxygen bond (1.45 Å) whereas for similar final positions in silicon (# in

Table III) O-O bond is stretched to 1.52 Å. Another evidence is the presence of stretched O-O physisorbed on two germanium atoms of two adjacent dimers (configuration 1, 8, 9) whereas in silicon a full dissociation is observed as shown in Figures 2 a and b respectively. Moreover, in silicon, no starting positions were able to bond adjacent dimers.

Table III. Summary of the oxidation of Si(100) surface extracted from Ref. 7. Starting positions are referred with regard of the starting positions described in Table I for Ge substrate and final structures are schematized. Adsorption energies are given in eV. White spheres represent oxygen atoms, grey ones represent silicon atoms. Grey spheres size and darkness vary as a function of the layer depth from the top: Larger the sphere is the topmost the layer.

Adsorption mechanism	Above the dimers row	Above or near a single dimer silicon atom Above the channel		
Starting Positions	2	1,9,13,14	3,4,5	6,7,8,10,11,12,15,16
Final Structure-types	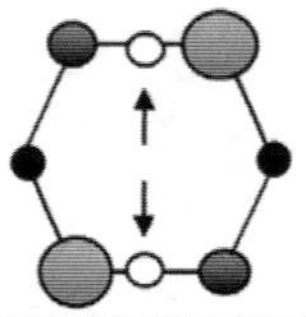	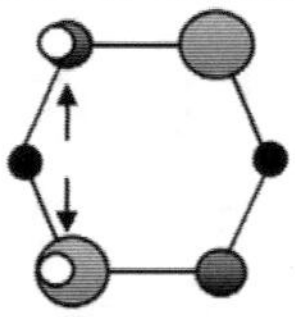	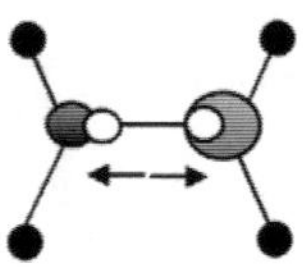	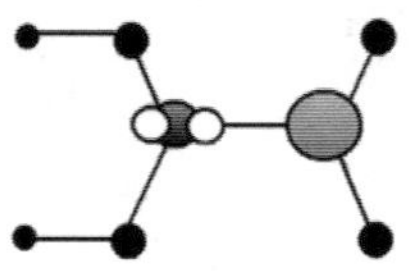
Eads (eV)	6.67	1.7 to 2.3	3.2	2 to 2.2
d(O-O) (Å)	Fully incorporated	Fully dissociated	1.52	1.59

In a preliminary conclusion, we unravel that different oxidation intermediates are found for initial oxidation stage on Si and Ge (on top; peroxy bridge, configuration respectively). In the next part, we perform a static study for evaluating oxygen atom ability to insert into germanium material.

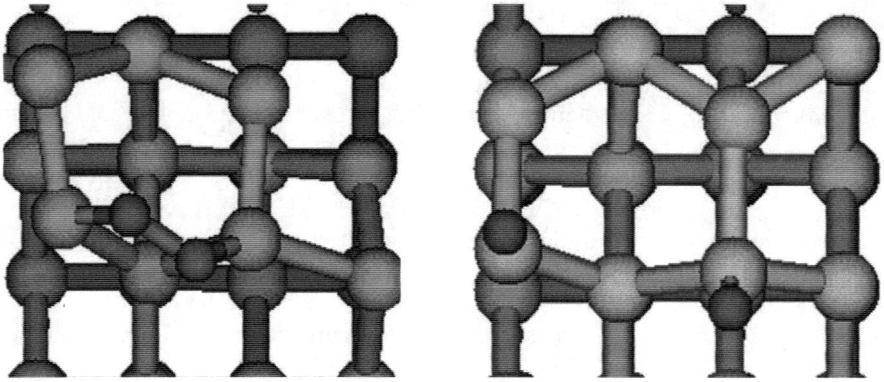

Figure 2. Top views of adsorbed configuration #9 on Germanium substrate (left picture) and on silicon substrate (right picture). Ge atoms are in grey, Si atoms are in yellow and hydrogen atoms are in white.

As a starting point, we consider a set of structures having one or two oxygen atoms attached and/or inserted into germanium surface. Germanium results will be compared to previous conclusions relative to the stability of Si and SiGe oxidized structures having one and two oxygen atoms incorporated in similar surface configurations. The structures and associated stability are listed respectively in the Tables IV and V. In each case, the 0 eV energy reference is taken as the most stable structure. Even if similar structures can be found in silicon and germanium slabs, one can see that their stability varies as a function of the substrate nature.
The insertion of one oxygen atom into the surface dimer is denoted as 'dimer insertion', while insertion into the bond between the first and the second layer is referred to as 'backbond insertion'. One oxygen atom not incorporated and stabilized on a single dimer atom is called 'on top' configuration and one in bridging position between two dimers is a 'bridging' configuration.

Table IV. Comparative energetic diagrams of one oxygen atom adsorbed a. on the Ge(100)-(2x1), b. on the Si(100)-(2x1) and c. on the SiGe(100)-(2x1) surfaces. Relative energies are given in eV. Top view surface structures are schematized. White spheres represent oxygen atoms, grey and dark spheres represent silicon atoms and grey and dark squares represent germanium atoms. Spheres and squares size vary as a function of the layer depth from a top point of view: Larger the sphere or square is the topmost the layer.

	A	B	C	D	E	F
a/ 10-atoms structures on Ge(100)-(2x1)						
	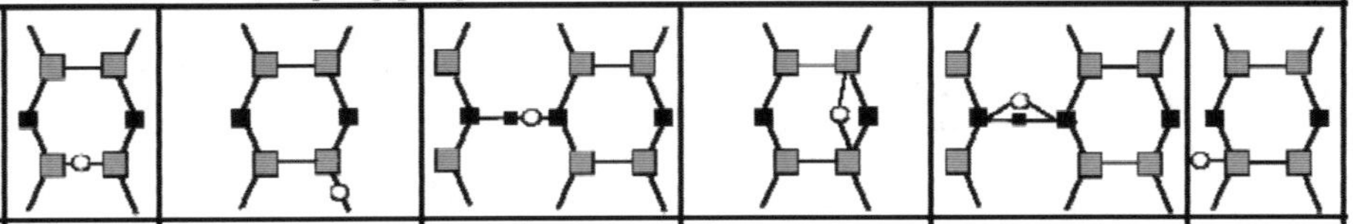					
ΔE (eV)	0	0.22	0.40	1.09	1.20	1.25
b/ 10-atoms structures on Si(100)-(2x1)						
	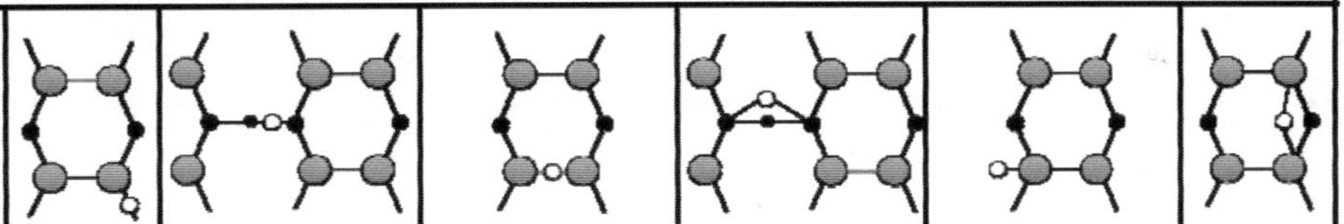					
ΔE (eV)	0	0.18	0.22	0.53	1.27	1.48
c/ 10-atoms structures on SiGe(100)-(2x1)						
	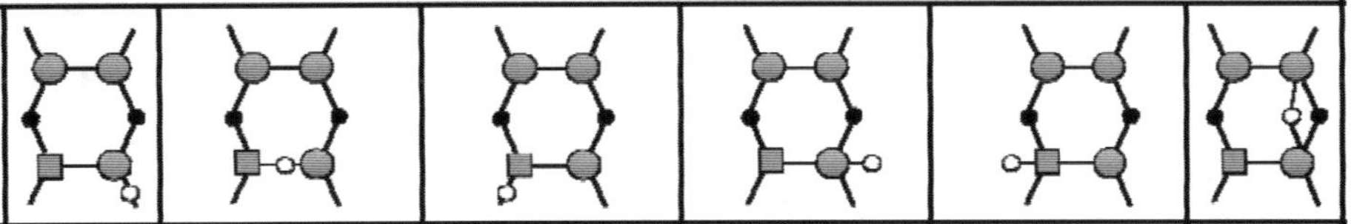					
ΔE (eV)	0	0.54	0.79	1.20	2.63	unstable

We first consider the case of one oxygen atom incorporated into the topmost layer (Table IV).
For all three substrates, one oxygen atom fully incorporated into Si-Si or Ge-Ge bond is the most favorable configuration that can be observed: oxygen atom in an 'on top' configurations is thus a metastable oxidation, mainly observed at the early stages of oxidation process and already known as an unavoidable step during silicon oxidation [7,9,12]. Insert into substrate is therefore a two steps process (thermodynamically more favorable). In the case of germanium material, dimer insertion is reached whereas in silicon, a backbond insertion is preferred. One can notice here that a reversed

IOP Conf. Series: Materials Science and Engineering **41** (2012) 012007 doi:10.1088/1757-899X/41/1/012007

stability order exists between these structures: in a germanium surface, a dimer insertion (Table IV.a-A) will be more stable by 0.22 eV than a backbond insertion (Table IV.a-B), compared to silicon, where a backbond configuration (Table IV.b-A) is more energetically favorable by 0.22 eV than dimer oxidized structure (Table IV.b-C). Interestingly, in the SiGe system, the oxidized Si-Si backbond is the most stable configuration, highly larger than a dimer insertion by 0.54 eV. The 'on top' configuration is less stable by roughly 1.25 eV than the energy reference structure (structure IV.a-F and structure IV.b-E for Ge and Si respectively).
Inter-dimer bridge structure given in IV.a-D is a clue that characterizes the softness of germanium surfaces: this structure (Table IV.a-D) is more stable than an on top configuration by 0.16 eV (Table IV.a-F) compared to silicon case, where a inter-dimer bridge structure (Table IV.a-F) is less stable by 0.21 eV than the on top configuration (structure IV.a-E). In silicon material, we see that an inter-dimer bridging state costs in energy and is a non-favorable step at the primary stages. This trend changes when more than two oxygen atoms are inserted in the surface due to charge transfers effects [13].

Table V. Comparative energetic diagrams of two oxygen atoms adsorbed a. on the Ge(100)-(2x1), b. on the Si(100)-(2x1) and c. on the SiGe(100)-(2x1) surfaces. Relative energies are given in eV. Top view surface structures are schematized. White spheres represent oxygen atoms, grey and dark spheres represent silicon atoms and grey and dark squares represent germanium atoms. Spheres and squares size vary as a function of the layer depth from a top point of view: Larger the sphere or square is the topmost the layer.

	A	B	C	D	E	F	G
a/ 2O-atoms structures on Ge(100)-(2x1)							
	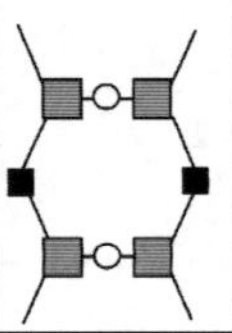	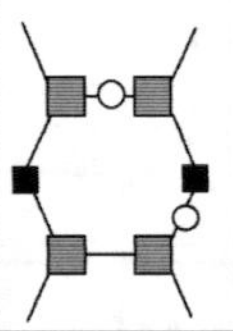	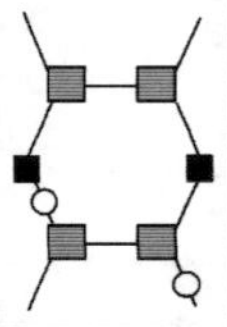	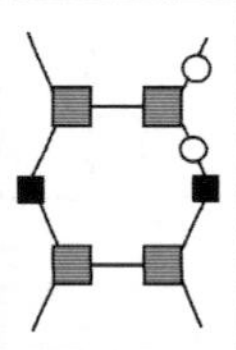	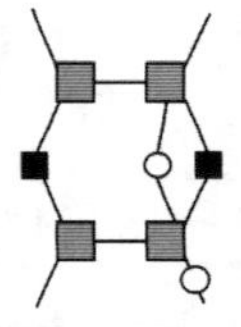	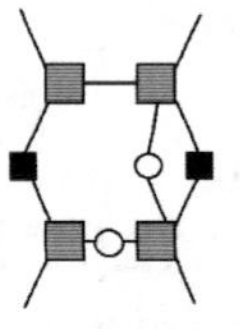	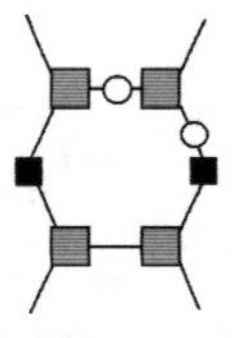
ΔE (eV)	**0**	**0.30**	**0.67**	**0.70**	**0.85**	**1.03**	**1.1**
b/ 2O-atoms structures on Si(100)-(2x1)							
	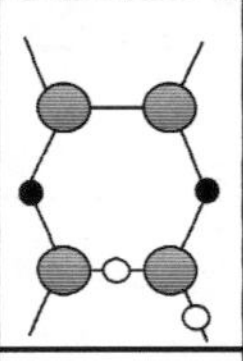	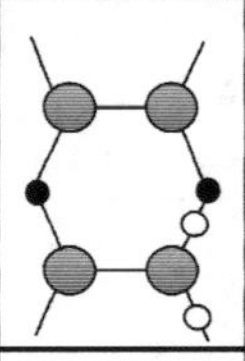	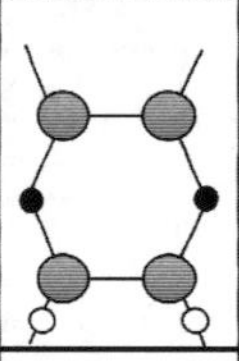	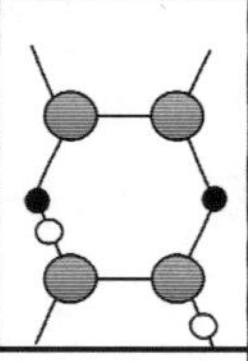	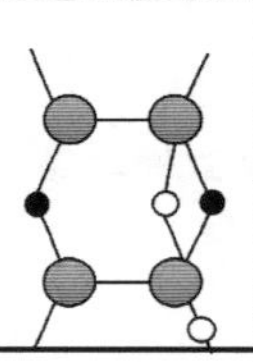	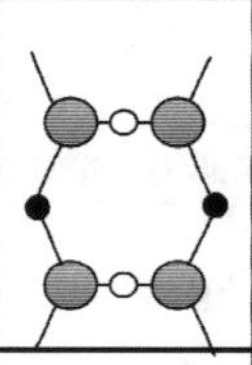	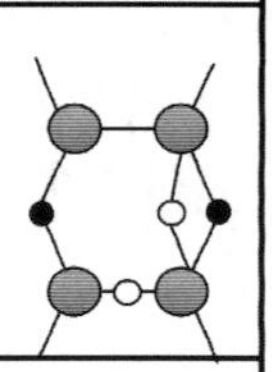
ΔE (eV)	**0**	**0.15**	**0.19**	**0.35**	**0.46**	**0.64**	**1.36**
c/ 2O-atoms structures on SiGe(100)-(2x1)							
	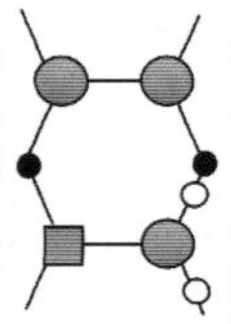	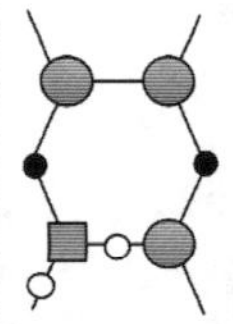	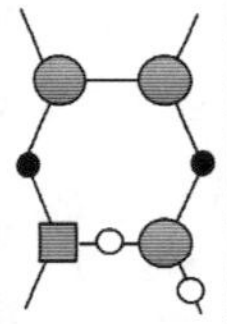	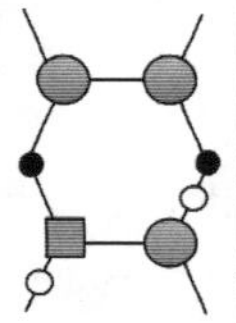	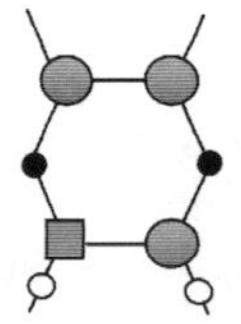		
ΔE (eV)	**0**	**0.29**	**0.39**	**0.65**	**0.66**	**/**	**/**

This first static study suggests an important role of the 'on top' configuration during the initial stage of both germanium and silicon oxidation processes. It also exhibits two different insertion pathways: germanium oxidation proceeds via oxygen atom incorporation into the dimer bond whereas silicon has a more complex oxidation phenomenon with a preferential incorporation site into the backbond but a kinetically favorable insertion into the dimer [2]. This behaviors are also is also illustrated in the SiGe case. For hybrid system, oxygen atom seems to avoid the close neighborhood of Ge atom and prefers to incorporate into Si-Si bond. Then, a mixed Ge-Si 'dimer insertion' (structure IV.c-B) is energetically better than a mixed Ge-Si 'backbond insertion' (structure 2c-D).

The most stable germanium configuration has oxygen atoms inserted into two adjacent dimers (Table V.a-A). In opposition, the most stable configuration observed in silicon surface is obtained on a single dimer unit with a more "compact" oxide structure: one oxygen atom in a backbond and one oxygen atom attached on top, this structure is known as the silanone structure and detailed can be found in Ref. 9 We observe in germanium substrate that it is tricky to adsorb two oxygen atoms in the vicinity of each other. Here again, an opposite remark can be derived from Ge and Si oxidation mechanisms: a structure with two oxygen atoms adsorbed on the same dimer unit (structure V.a-C) is less stable by 0.67 eV than the two oxidized dimer units structure. In silicon material, it costs 0.64 eV to adsorb two oxygen atoms, each on adjacent dimers compared to the silanone configuration.
Here, the propensity of oxygen atoms to agglomerate on silicon surface contrasts with germanium material where oxygen atoms seem to repeal each other. In his publication Soon [14] has proposed different reaction pathways for the decomposition of one oxygen molecule one a germanium cluster. In this study the author connects two initial configurations presented in the Table 1 (configuration 2 and 3) to structures presented in the table V.a. He shows that structure V.a-C and structure V.a-D are resulting from the adsorption and decomposition of one oxygen molecule centered between two dimers and perpendicular to the dimer row (configuration 2) via the structure V.a-G. It also connects the structure V.a-B to the adsorption and decomposition of one oxygen molecule initially above and parallel to one dimers unit (configuration 3) via the structure V.a-A. Even if the stability order of the structure presented in this paper is different to those presented in our present paper, probably due to bordering interactions in the cluster approach used to simulate the surface, the considerable activation barrier found by Soon to take on oxygen atom in the dimer unit and to insert it in the backbond support the idea of a preferential oxidation of dimer.

5. Conclusion

In this paper, the first stages of germanium oxidation are studied by DFT calculations and a systematic comparison with silicon and silicon germanium data is done. Both oxygen in on-top and peroxy-bridge configurations appear as intermediate metastable unavoidable steps in germanium oxidation process. Peroxy bridge formation is allowed thanks to the softness of the germanium surface giving the possibility to germanium atoms to move out from their crystalline position to adapt to the approaching oxygen molecule. Dissociation is observed in silicon where inter-dimer bridging configurations occur in germanium. So, in pre-oxidation regime Germanium material, germanium atoms adapt to the oxygen species, whereas in silicon material, oxygen molecule adapts to the silicon surface topology. The oxygen atoms tendency to agglomerate on the surface is described in opposition to the favorable spreading of isolated bridging oxygen configurations on germanium surface.
These elemental but fundamental differences in behavior of both silicon and germanium surfaces towards oxidation is key to unravel the dramatic differences observed in providing high quality semiconductor oxide interfaces with these materials. On going research effort is engaged to shad light into the next oxidation stages.

IOP Conf. Series: Materials Science and Engineering **41** (2012) 012007 doi:10.1088/1757-899X/41/1/012007

5. References

[1] Dimoulas A et al., Appl. Phys. Lett. 86 (2005) 032908.
[2] Hemeryck A. et al., Surf. Sci. 601 (2007) 2339.
[3] Dkhissi A. et al., Appl. Phys. Lett. 94 (2009) 041912.
[4] Kresse G et al., Phys. Rev. B 54 (1996) 11169.
[5] Kresse G et al., Comput. Mater. Sci. 6 (1996) 5.
[6] Monkhorst H J et al., Phys. Rev. B 13 (1976) 5188.
[7] Hemeryck A, et al., J. of Non-Cryst. Sol. 353 (2007) 594.
[8] Soon J M, et al., Phys. Rev. B 72 (2005) 115343.
[9] Hemeryck A, et al., J. Chem. Phys. 126 (2007) 114707.
[10] Richard N. et al., Comp. Mater. Sci. 33 (2005) 26.
[11] Fan X L. et al., J. Phys. Chem. C 113 (2009) 8786.
[12] Hemeryck A. et al., Surf Sci 601 (2007) 2339.
[13] Hemeryck A. et al., Phys. Rev. B 79 (2009) 035317
[14] Soon J M., et al., Phys. Rev. B 72 (2005) 115343.

6. Acknowledgments

This work was performed using HPC resources from CALMIP and by LAAS-CNRS resources.

Electron Trap Energy Distribution in ALD Al_2O_3, $LaAl_4O_x$, and $Gd_yAl_{2-y}O_3$ Layers on Silicon

W. C. Wang[1], M. Badylevich[1, 2], C. Adelmann[3], J. Swerts[3], J. A. Kittl[3] and V.V. Afanas'ev[1]

1. Department of Physics, University of Leuven, B-3001 Leuven, Belgium
2. Present affiliation: Light and Lighting Laboratory, Catholic University College Gent, Belgium
3. IMEC, B-3001 Leuven, Belgium

E-mail: WanChih.Wang@fys.kuleuven.be

Abstract. The energy distribution of electron trap density in atomic layer deposited Al_2O_3, $LaAl_4O_x$ and $Gd_yAl_{2-y}O_3$ insulating layers was studied by using the exhaustive photodepopulation spectroscopy. Upon filling the traps by electron tunneling from Si substrate, a broad energy distribution of trap levels in the energy range 2-4 eV is found in all studied insulators with trap densities in the range of 10^{12} $cm^{-2}eV^{-1}$. The incorporation of La and Gd cations reduces the trap density in aluminate layers as compared to Al_2O_3. Crystallization of the insulator by the post-deposition annealing is found to increase the trap density while the energy distribution remains unchanged. The similar trap spectra in the Al_2O_3 and La or Gd aluminate layers suggest the common nature of the traps, probably originating from imperfections in the AlO_x sub-network.

1. Introduction

The wide bandgap (6.2 eV for amorphous phase[1] and 8.7 eV for cubic γ-phase[2]) of high-κ aluminum oxide Al_2O_3 (κ~9) has made it attractive to be applied as the interpoly dielectric in the embedded Flash[3] and as the blocking insulator in TANOS device[4]. The high permittivity of Al_2O_3, as compared to SiO_2 (κ=3.9), allows one to reach a lower effective oxide thickness and in this way to reduce the operation voltages. However, the electron trapping in Al_2O_3 layer hampers ERASE by producing a fixed negative charge. The knowledge of the trap density and the trap energy distribution are essential to solve the problem. The rare-earth-based lanthanum aluminate $LaAl_4O_x$ (κ~10-16) and gadolinium aluminate $Gd_yAl_{2-y}O_3$ (κ~12-17[5]) are also investigated here for their higher κ value to evaluate sensitivity of these insulators to electron trapping.

In this work, the electrons are found to be easily trapped in the Al_2O_3, $LaAl_4O_x$, and $Gd_yAl_{2-y}O_3$ insulators. A broad trap energy depth (E_t) distribution is found in all the studied layers, instead of the discrete defect level expected in a crystalline phase. The similar deep traps distribution from 2 to 4 eV below the conduction band (CB) bottom in samples with both amorphous and crystallized insulator is suggested to be related to the AlO_x subnetwork encountered in all oxides. Replacement of some Al cations by the La or Gd reduces the trap density to around one half compared to the trap density in Al_2O_3 layer. However, considerable density of shallow traps (< 2 eV) are found in $LaAl_4O_x$ while no such effect is seen in $Gd_yAl_{2-y}O_3$ layers.

2. Experimental

2.1. Sample Preparation

Studied samples were prepared on p- or n-type (n_a, $n_d \approx 10^{15}$ cm^{-3}) (100)Si wafers covered by a 4-nm thick thermally grown tunnelling SiO_2 layer. On the top of SiO_2 layer, the Al_2O_3, $LaAl_4O_x$ and $Gd_yAl_{2-y}O_3$ layers were grown by atomic layer deposition (ALD) technique. A 12-nm thick amorphous Al_2O_3 (a-Al_2O_3) layer was deposited from trimethyl-aluminum (TMA) and ozone (O_3) precursors at 300 °C. Then some of the Al_2O_3 samples were annealed in N_2 ambient at 1000 °C for 1 min, resulting in polycrystalline cubic (γ-phase) Al_2O_3 layers. To deposit a 20-nm thick $LaAl_4O_x$ layer, La(thd)$_3$, TMA and O_3 were used as the ALD precursors at 300 °C. Some of the $LaAl_4O_x$ samples were also annealed in N_2 ambient at 800 °C for 1 min, leading to the oxide crystallization in cubic phase[6]. The 18-nm thick $Gd_yAl_{2-y}O_3$ layers with different Gd content were deposited using Gd(^{i}PrCp)$_3$, TMA and H_2O precursors at 250 °C, and were annealed in N_2 ambient at 700 °C for 1 min[5]. Finally, 15-nm thick Au electrodes of 0.5 mm^2 area were thermoresistively evaporated through the shadow mask on insulators, forming a metal-oxide-silicon (MOS) capacitor structure.

2.2. Measurement Procedure

Electron traps in SiO_2/insulator stack were filled by the electron injection from the Si substrate induced by applying a sufficient positive bias (V_g) to the Au metal gate and simultaneously illuminating the sample by the infrared light (photon energy hν<1.42 eV) to provide sufficient concentration of electrons at the Si surface. After the electron injection, the sample was kept in darkness for around 10 hours to reach reproducible capacitance-voltage (C-V) traces. The photodepopulation experiments were performed at room temperature in the hν range 1.3-6.1 eV, with the hν increasing in steps of Δhν=0.3 eV after exposing the sample to the light for one hour to ensure saturation of the electron de-trapping kinetics[7]. After each illumination step, the capacitor flatband voltage shift (ΔV_{FB}) was monitored by measuring the 1 MHz C-V curves.

Variations in the trapped charge density during illumination occur due to predominantly the optical excitation of electrons from the traps[7]. The excited electrons drift away from the insulator for the external bias V_g=1 V is applied on the top Au electrode during the photodepopulation experiment. The spectral charge density (SCD) relative to the insulator CB bottom can be found from the measured shift of the C-V curve using equation[7]:

$$\frac{\Delta Q(h\nu)}{\Delta h\nu \cdot q} = -2\left[\frac{\varepsilon_0 \varepsilon_{SiO_2}}{\varepsilon_0 \varepsilon_{SiO_2} - d_{SiO_2} \cdot C}\right]^2 C \cdot \Delta V_{FB}(h\nu) \qquad (1)$$

, where ΔQ(hν) is the variation of the charge in insulator in the specific hν interval, C is the specific insulator capacitance, $\varepsilon_0 = 8.854 \times 10^{-12}\ F/m$ is the vacuum permittivity, $\varepsilon_{SiO_2} = 3.9$ is the silicon dioxide dielectric constant, and d_{SiO_2} is the thickness of the tunneling SiO_2 layer.

3. Results and Discussion

A decrease of the negative charge density upon illumination is observed during the photodepopulation experiments in all the studied samples, corresponding to the electron removal from the insulator layer or, else, to the hole capture injected from the metal gate. To distinguish between these two mechanisms of the trapped electrons removal, we repeated the measurements using fresh samples, not subjected to electron injection. Since we found no net positive charge trapping expected to occur should holes be injected into the insulating stacks[8], the photo-excitation of electrons from the trap levels must be the dominant charge-variation process during illumination.

The energy distribution of electron traps in Al_2O_3, La- and Gd-aluminates layers will be discussed in the following sections. The SCD *versus* hν plots as measured in the electron-injected samples are shown by open bars (□) in figures 1-3. The uncharged samples were studied for comparison to evaluate the energy depth distribution and the density of the traps already filled during the sample preparation process, shown by dashed bars (▨) in figures 1-3.

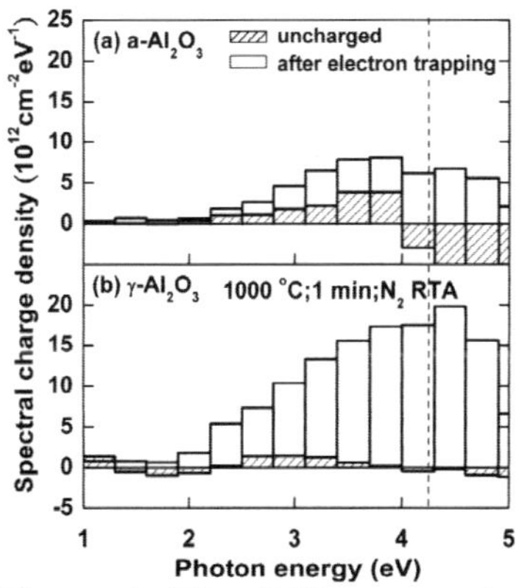

Figure 1. The SCD *versus* hν plots of (a) amorphous Al_2O_3 and (b) crystallized γ-Al_2O_3 layers.

3.1. Al_2O_3

The SCD as a function of hν in the samples with a-Al_2O_3 insulator is plotted in figure 1(a). Almost no measureable charge variation is seen at hν<2 eV, suggesting the absence of electron states available for photoexcitation in this energy range. In the range 2<hν<4 eV, the charge of the trapped electrons is removed upon illumination, suggesting photodepopulation of the traps. The amount of de-trapped charge in this spectral range clearly correlates with the density of initially trapped electrons supporting the latter conclusion. In the uncharged sample a small fraction of traps appears to be already occupied, probably due to the electron tunneling from Si during Al_2O_3 layer deposition. Also, as already mentioned above, no net positive charge is observed, which means there is no significant influence of hole injection from Au electrode on the observed SCD distributions.

When hν exceeds 4 eV, the electron de-trapping rate in the electron-injected sample decreases while the net electron trapping is observed in the fresh (uninjected) sample. This may be explained by photoinjection of electrons from silicon substrate surmounting the Si/SiO_2 interfacial barrier; the zero-field (100)Si/SiO_2 barrier height of 4.25 eV[9] is marked by the vertical dashed lines in figures 1-3. Above this spectral threshold electrons are photoinjected from the valence band (VB) of silicon into SiO_2 and may be trapped in Al_2O_3 layer. As a result, the SCD can be considered as the representative of the trapped electrons energy distribution only in the energy range hν<4 eV.

Crystallization of alumina has profound effect on the electronic structure of the oxide by increasing its gap to around 8.7 eV. This gap widening occurs in a highly asymmetric way by a ~ 0.5 eV shift upward for the CB bottom and a ~2 eV shift downward for the VB[1]. The electron traps also appear to be influenced strongly: The SCD curves shown for the crystallized sample in figure 1(b) indicate the at least doubling of the trap density as compared to a-Al_2O_3 [panel (a)]. Moreover, the electric field of trapped electrons appears to be sufficient to block the photoinjection from silicon thus allowing us to trace the energy distribution of trapped electrons down to nearly 5 eV below the γ-Al_2O_3 CB edge. In both a-Al_2O_3 and γ-Al_2O_3 cases one observes a broad distribution of the trapped electron energies and, remarkably, the overall shape of the trap distribution below 4 eV remains unchanged in a- and γ-Al_2O_3 samples, suggesting a common origin of the traps. Moreover, as the E_t distribution is unaffected by the shift of the oxide CB bottom, the energy levels of trapped electrons are probably generically derived from the oxide CB electron states and shift in accordance with the latter during Al_2O_3 crystallization. This observation would suggest the traps to be related to aluminum network, as the bottom of CB in Al_2O_3 is derived from the unoccupied electron states of Al cations.

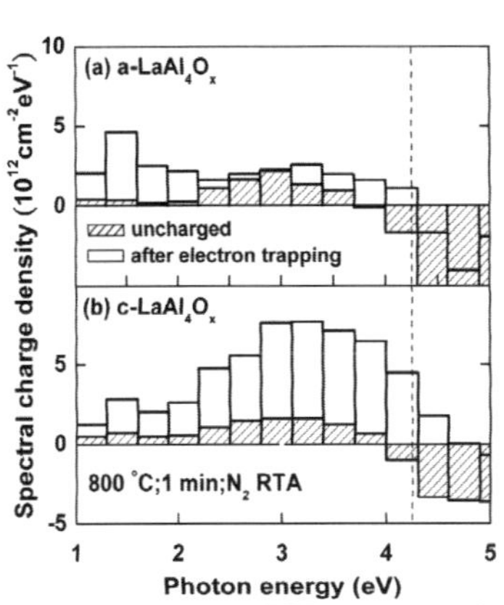

Figure 2. The SCD *versus* hν plots of (a) amorphous $LaAl_4O_x$ and (b) crystallized $LaAl_4O_x$ layers.

3.2. $LaAl_4O_x$

The SCD energy distributions for as-deposited a-$LaAl_4O_x$ and crystallized $LaAl_4O_x$ (c-$LaAl_4O_x$) films are compared in figure 2(a), and (b), respectively. Unlike the Al_2O_3 layers, shallow traps with the energy depth smaller than 2 eV are found both in a- and c-$LaAl_4Ox$ samples. The density of these traps is slightly reduced after annealing-induced crystallization of the $LaAl_4O_x$ layers. As compared to Al_2O_3 layers, the lower trap densities in $LaAl_4O_x$ layers in the energy depth interval 2<hν<4 eV suggests that shallow traps may assist de-trapping during high-field electron injection pulse. Interestingly, the SCD distribution of deep traps (2<hν<4 eV) remains similar to that in Al_2O_3 layers. Furthermore, after crystallization the trap density in this energy range more than doubles in the trend which is similar to that in Al_2O_3 (figure 1). Then, these deep traps in 2<hν<4 eV may originate from the same defects in AlO_x subnetwork. In their turn, shallow traps might be related to LaO_x network as they are not found in the La-free insulators.

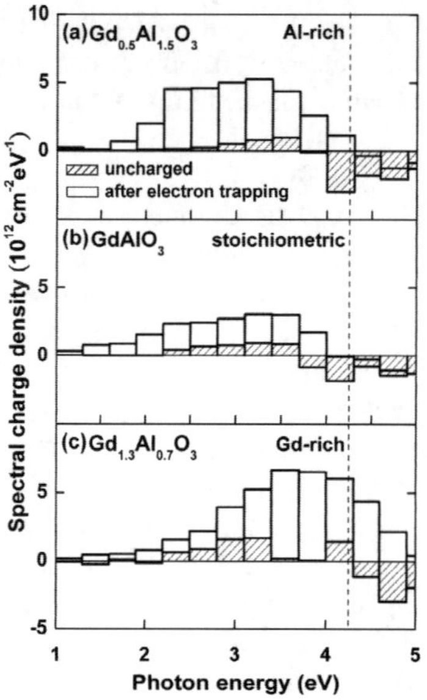

Figure 3. The SCD *versus* hν plots of (a) Al-rich, (b) stoichiometic, and (c) Gd-rich annealed Gd-aluminates.

3.3. $Gd_yAl_{2-y}O_3$

To get more insight on the nature of the electron traps, we compared SCD distributions in the annealed Gd-aluminate layers of different stoichiometry, as shown in figure 3. Notably, no high density of the shallow traps is found in the Gd-aluminates as opposed to the $LaAl_4O_x$ case. The electron trap distributions in the Al-rich and stoichiometric Gd-aluminate layer in the range of 2<hν<4 eV are similar to those found in the Al_2O_3 layers. By contrast, the Gd-rich $Gd_{1.3}Al_{0.7}O_3$ layer shows that most of the traps have energy depth exceeding 3 eV. Notably, the trap density in the stoichiometric $GdAlO_3$ film is the lowest, comparable to that in a-$LaAl_4O_x$ layer. Overall, the $Gd_yAl_{2-y}O_x$ samples have less traps as compared to Al_2O_3, which, taking into account the higher permittivity of this material makes it one of the best choices for an interpoly insulator.

4. Conclusions

As the most striking result of the present study one may indicate a broad energy distribution of electron traps is found in ALD Al_2O_3, $LaAl_4O_x$ and $Gd_yAl_{2-y}O_3$ layers, which is inconsistent with the hypothetical discrete energy level of defects expected in single-crystal materials. Generally, the deep traps are distributed in 2 to 4 eV range in both as-deposited and crystallized samples suggesting the same defect states to be involved, and pointing towards some Al-related entity.

Interestingly, crystallization of the oxide by annealing effectively increases the trap density without affecting much the SCD spectra. The trap density is seen to be influenced by the oxide composition without changing significantly the energy distribution as found in Gd-aluminates. As it is revealed by our experiments, by introducing the rare earth cation La or Gd one not only increases the κ value of the insulating layer but also decreases the trap density, suggesting rare-earth aluminates as potential candidates for the blocking insulator in FLASH cells.

References

[1] Afanas'ev V V *et al* 2011 Appl. Phys. Lett. **99** 072103
[2] Ealet B *et al* **250** 92
[3] Zheng X F *et al* 2010 IEEE T. Electron. Dev. **57** 288
[4] Zahid M B *et al* 2010 IEEE T. Electron Dev. **57** 2907
[5] Adelmann C *et al* 2010 Chem. Vapor Depos. **16** 170
[6] Swerts J *et al* 2011 Appl. Phys. Lett. **98** 102904
[7] Wang W C *et al* 2009, Appl. Phys. Lett. **95** 132903
[8] Afanas'ev V V, and Stesmans A 2004 J. Appl. Phys. **95** 2518
[9] Adamchuk V K, and Afanas'ev V V 1992 Prog. Surf. Sci. **41** 111

Gigahertz monolithic delay lines for surface acoustic waves on Silicon

P. V. Santos[1,3], S. Rauwerdink[1], K. Biermann[1], B. Drescher[1], W. Seidel[1], M. Kaynak[2], U. Kaletta[2], M. Fraschke[2], D. Wolansky[2], and Ch. Wenger[2]

[1] Paul-Drude-Institut fuer Festkoerperelektronik, Hausvogteiplatz 5–7, D-10117 Berlin, Germany
[2] IHP, Im Technologiepark 25, 15236 Frankfurt (Oder), Germany

[3] E-mail: santos@pdi-berlin.de

Abstract. We demonstrate the electrical generation of GHz surface acoustic waves (SAWs) on silicon chips using a process entirely compatible with standard CMOS fabrication. In this process, the interdigitated acoustic transducers are embedded in a silicon dioxide layer, which is subsequently coated with a piezoelectric ZnO film. We show that delay lines with sub-µm wavelength fabricated using this process reach resonance frequencies approaching 4 GHz.

1. Introduction

Surface acoustic wave (SAW) devices find applications in a wide field including signal processing, light modulation and different kinds of sensors. In most cases, these devices are fabricated on piezoelectric material in order to allow for the electrical generation of the SAWs using interdigitated acoustic transducers (IDTs). Since silicon is not piezoelectric, acoustic devices are normally fabricated in a hybrid configuration with the piezoelectric material electrically connected to the external driving and control electronics. The monolithic fabrication of acoustic elements on silicon chips becomes possible via the deposition of thin piezoelectric films on the silicon surface. In addition to more compact devices, monolithic fabrication can also take advantage of the high photolithographic resolution of standard silicon processing to produce IDTs for high-frequency applications. In many cases, however, full integration is hindered by the incompatibility of piezoelectric materials (ZnO, for instance) with silicon processing. Even though different approaches have been presented, most of them have been restricted to the fabrication of SAW elements for relatively low frequencies (below 1 GHz) [1-6].

In this letter, we demonstrate that the integration of SAW elements in a conventional Si CMOS process can be carried out by taking full advantage of the high photo-lithographic resolution to produce GHz SAW devices. The integration approach is displayed schematically in Fig. 1(a). Instead of the conventional design with the transducers placed on top of the piezoelectric film, the IDTs in our case consist of tungsten (W) stripes embedded in an amorphous SiO2 (a-SiO2) layer underneath a piezoelectric ZnO film. A similar embedded approach has been used to fabricate low-frequency IDTs on Si [2–4, 6] and $LiNbO_3$ [7, 8] substrates. The high photolithographic resolution of the CMOS process allows for the fabrication of acoustic delay lines for frequencies approaching 4 GHz. The resonance frequencies of the modes generated by the embedded IDTs is well-explained by an elastic model for the layer structure.

2. Experimental details

The experiments were carried out on delay lines consisting of two IDTs oriented along a x = ⟨110⟩ surface direction of z = (001) silicon wafers (cf. Fig. 1(b)). The IDTs have aperture and length of 31.5 and 36 μm, respectively, and are separated by a 63 μm-long SAW propagation path. We have fabricated delay lines using IDTs with W finger widths ranging from 190 to 600 nm, which cover acoustic wavelengths λ_{SAW} from 0.76 to 2.4 μm. The scanning electron micrograph of Fig. 1(c) displays the cross-section of a test structure showing the W fingers embedded in a 400 nm a-SiO_2-layer placed underneath the piezoelectric ZnO film. For the IDT fabrication, 180 nm-thick W-fingers were embedded into an a-SiO_2 layer using a W chemical vapor deposition process optimized to fill grooves etched on the SiO_2 surface using a C_2F_4-based reactive ion etching. The sample surface was then planarized by Chemical Mechanical Polishing (CMP), yielding a residual surface height modulation (between the W-coated and uncoated areas) of only 20 nm, which is considerably smaller than the SAW wavelength. The thickness of the a-SiO2 layers in the IDTs for $\lambda_{SAW} < 1.8$ μm and $\lambda_{SAW} \geq 1.8$ μm were 400 and 600 nm, respectively.

The previous IDT fabrication steps, which require high-resolution lithography, were carried out in a standard CMOS environment. The 500 nm-thick ZnO film was subsequently deposited at 200 °C using a sputtering process optimized to yield columnar films with the wurzite c-axis along the growth direction. The orientation and, therefore, the piezoelectric properties of this layer depend critically on the nucleation conditions on the W-grating surface. In order to ensure reproducible nucleation conditions, the samples were briefly sputter-etched in an Argon environment prior to the ZnO deposition. In order to have a uniform surface for the nucleation of the ZnO crystallites, a thin (50 nm-thick) a-SiO_2 buffer layer can be sputtered on the W-grating, as shown in Fig. 1(c). We found, however, that this buffer layer is not required for the well-planarized surfaces produced by the CMP process – this buffer was also not used in the samples studied here. In the final fabrication step, the ZnO film was etched from the pad areas to allow access to the W-contacts. The fabrication sequence separates the CMOS processes from the deposition of the piezoelectric layers: it avoids, therefore, contamination effects and can be easily extended to other piezoelectric materials.

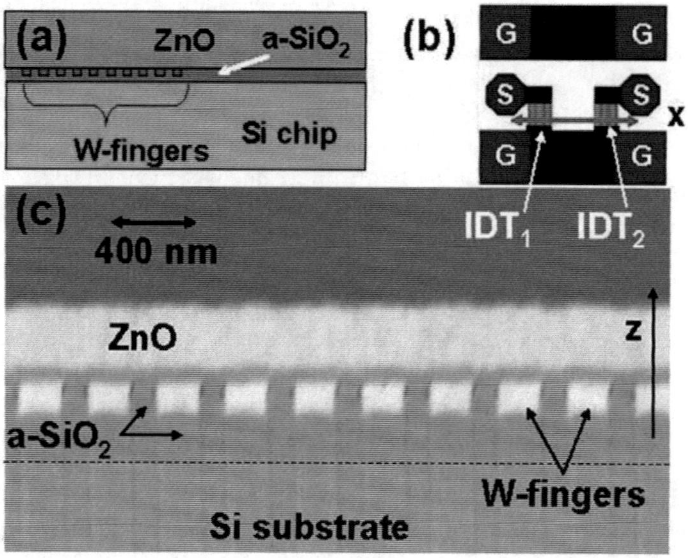

Figure 1. (a) Structure of embedded IDTs on a silicon chip and (b) Top view of the tungsten (W) layer underneath the ZnO film in a SAW delay line, showing the signal (S) and ground (G) rf-contact pads. The arrow displays the 150 μm-long scan line for Fig. 5. (c) Cross-sectional scanning electron micrograph of a test IDT with a 50 nm-thick a-SiO_2 layer between the IDT and the ZnO film.

3. Results

Figure 2 compares the rf-reflection parameter s_{11} of delay lines for different λ_{SAW} consisting of single-finger (SF) IDTs with finger widths equal to $\lambda_{SAW}/4$. Each delay line displays two resonances (denoted by A and B and marked by the dashed and solid arrows, respectively), which are attributed to modes with different depth profiles for the acoustic fields (see below). The s_{11} dips are very small for the long wavelength A-modes: these resonances, however, can be clearly distinguished in the s_{12} rf-transmission curves (not shown).

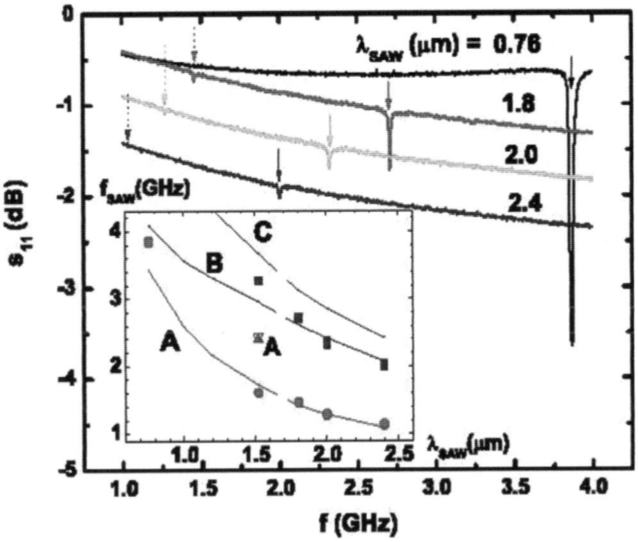

Figure 2:. Power reflection coefficient s_{11} of single finger IDTs with different SAW wavelengths (λ_{SAW}). The curves for λ_{SAW} = 2 and 2.4 μm are displaced vertically by 0.5 and 1 dB, respectively. The inset shows the dependence of the SAW resonance frequency (f_{SAW}) on λ_{SAW} for the A and B modes (symbols), which are indicated in the main plot by dashed and solid arrows, respectively. The triangles show data for the A' overtone observed in split-finger IDTs (cf. Fig. 3). The solid curves reproduce the calculated frequencies of the A, B, and C Rayleigh modes.

Figure 3 displays the reflection (s_{11} and s_{22}) as well as the transmission power coefficient (s_{12}) for a λ_{SAW} =1.52 μm delay line with split-finger (DF) IDTs, which contain 4 190 nm-wide fingers per SAW wavelength. In addition to the A and B modes, this delay line shows an additional resonance indicated by A'. For both types of delay lines, the resonance frequencies increase with decreasing λ_{SAW}, as illustrated in the inset of Fig. 2. By reducing λ_{SAW} below 1.8 μm, it becomes possible to reach frequencies above 3 GHz (frequencies of 3.25 and 3.8 GHz were obtained for λ_{SAW} =1.52 and 0.76 μm, respectively). Finally, by taking into account the short length of the IDTs and the fact that the delay lines were not optimized for electrical impedance matching, the relatively large s_{11} dips and s_{12} transmission peaks in Fig. 3 prove that acoustic absorption in the polycrystalline ZnO films does not hinder their application in GHz SAW devices.

Due to the layer structure of the samples, the SAW resonance frequencies do not scale inversely proportional to λ_{SAW} (inset of Fig. 2). In addition, the delay lines show an unusual electrical response with stronger electro-acoustic coupling for the higher overtones. In order to understand this behavior, we have calculated the frequency and spatial distribution of the acoustic field for the different modes following the procedure delineated in Ref. 9. In the calculations, the W-gratings were modeled as a continuous layer with elastic constants and density averaged between those of W and a-SiO_2. This approximation neglects, therefore, the effects of acoustic reflections at the individual fingers on SAW propagation.

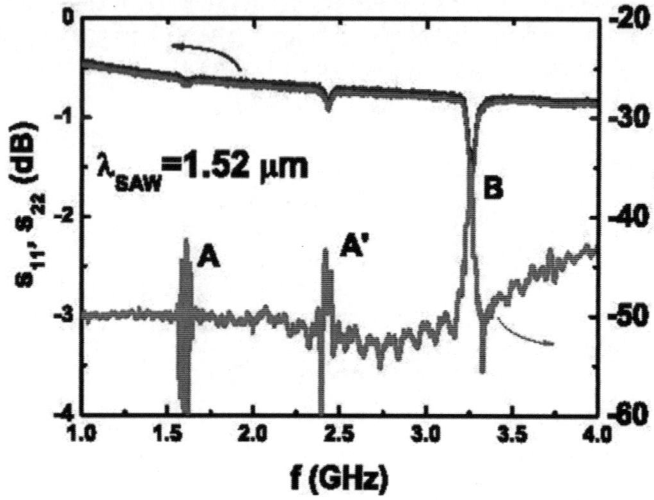

Figure 3. (a) Scattering parameters s_{11}, s_{22}, and s_{12}, and for a split-finger IDT for λ_{SAW} = 1.52 µm. A, A' and B denote the overtone series (cf. Fig. 2).

For each λ_{SAW}, the calculations predict a series of piezoelectric Rayleigh modes as well as non-piezoelectric Love modes. The inset of Fig. 2 compares the calculated frequencies for the lowest Rayleigh modes (the solid lines marked as A, B and C) with the experimental values. The calculations reproduce well the frequency of the A and B modes. Note, however, that the C modes have not been detected in the experiments, while the frequency of the A' mode is not reproduced by the calculations. These discrepancies are probably due to the increasing influence of acoustic reflections at the W-finger grating on the SAW propagation [10], which were neglected in the calculations. These effects are less pronounced in DF IDTs (cf. Fig. 3): as a result, they normally show wider s_{11} dips and higher rf-transmission than SF IDTs.

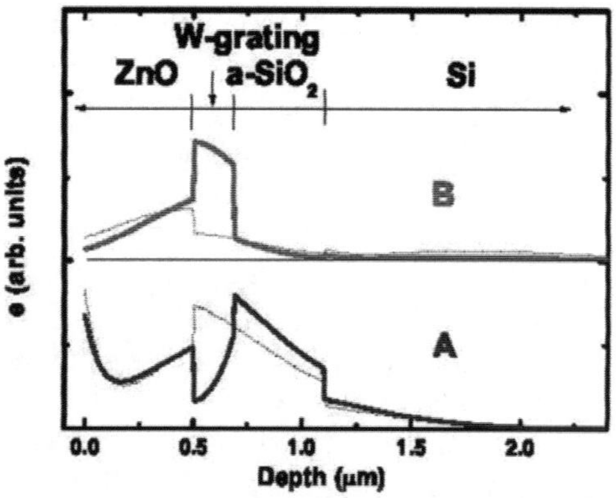

Figure 4. Calculated depth profiles for the acoustic energy density e for the A and B Rayleigh modes of a λ_{SAW} =2.4 µm SAW in structures with (solid lines) and without (dashed lines) a W-grating.

IOP Conf. Series: Materials Science and Engineering **41** (2012) 012009 doi:10.1088/1757-899X/41/1/012009

In addition to the resonance frequencies, the calculations also yields spatial depth profiles for the acoustic energy density (e), which are illustrated in Fig. 4 for modes A and B of a λ_{SAW} =2.4 µm SAW (solid curves). For comparison, the dashed lines show the corresponding profiles in the region between the IDTs, which were determined by replacing the W-grating by a homogeneous a-SiO_2 layer. In all cases, the acoustic energy concentrates in the overlayers with small penetration into the Si substrate. The overlayers behave, therefore, as the core of an acoustic waveguide with an effective SAW velocity lower than the one of the substrate. The acoustic energy concentration close to the ZnO film and W-grating is stronger for the B than for the A mode, thus providing a mechanism for the stronger electrical response of the higher frequency B-modes observed in Figs. 2 and 3.

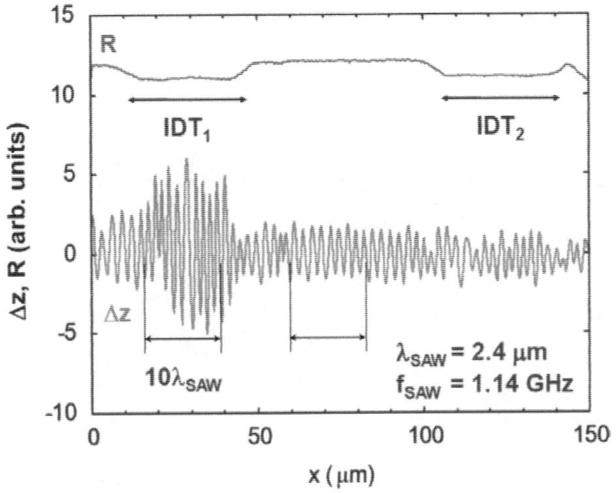

Figure 5. Optical reflection (R) and vertical surface displacement (Δz) recorded by interferometry along the scan line indicated in Fig. 1(b) for the A mode with λ_{SAW} =2.4 µm. The two sets of vertical bars are separated by 10 λ_{SAW} = 24 µm and show that the local wavelength increases outside the powered IDT (IDT1).

Due to the high W mass density, the W-grating has a strong impact on both the frequency and depth distribution of the SAW fields. The low conductivity of W (approx. 0.5% of that of Al) prevents the use of thin (i.e., $\ll \lambda_{SAW}$) W-fingers for the IDTs to minimize acoustic scattering. In addition, the W-grating reduces the acoustic velocity, even for the longest SAW wavelength studied. This effect can be observed by measuring the amplitude of the acoustic field along the delay lines using microscopic interferometry -details of the procedure are described in Ref. 9. Figure 5 shows profiles for the vertical surface displacement Δz for a fixed SAW phase recorded along the arrow indicated in Fig. 1(b). The upper curve shows the optical reflection (R), which decreases at the location of the two IDTs. In the measurements, the acoustic power was only applied to IDT1 (IDT2 remained unconnected). The amplitude of the Δz oscillations, which have the SAW periodicity, remains constant outside IDT1, thus attesting to negligible SAW absorption. Note, however, that the period of the oscillations is smaller within IDT1 than outside, thus indicating a lower acoustic velocity (by approx. 11±3%) in the W-grating region. This velocity reduction compares reasonably well with the value of 12% calculated by the model described above.

4. Conclusion

In conclusion, we have demonstrated the generation of SAWs using a process compatible with standard CMOS fabrication, thus opening the way for the monolitic integration of GHz acoustic devices on silicon chips.

5. References

[1] M. E. Motamedi, M. K. Kilcoyne, and R. K. Asatourian, IEEE Trans. Sonics Ultrasonics **SU-32**, 663 (1985).
[2] M. J. Vellekoop, E. Nieuwkoop, J. C. Haartsen, and A. Venema, IEEE Ultrasonics Symp. **87**, 641 (1987).
[3] J. H. Visser, M. J. Vellekoop, A. Venema, E. V. der Drift, P. J. M. Rek, A. J. Nederlof, and M. S. Nieuwenhuizen, IEEE Ultrasonic Symp. **1989** 195 (1989).
[4] P. T. M. van Zeijl, J. H. Visser, and L. K. Nanver, IEEE Trans. Cons. Electr. **35**, 512 (1989).
[5] M. J. Vellekoop, G. W. Lubking, P. M. Sarro, and A. Venema, Sensors and Actuators **A 44**, 249 (1994).
[6] D. M. Martin, V. Yantchev, and I. Katardjiev, J. Micromech. Microeng. **16**, 1869 (2006).
[7] M. Kadota and T. Kimura, Jpn. J. Appl. Phys. **45**, 4647 (2006).
[8] D. Reitz, Ph.D. thesis, Technischen Universitaet Dresden, Germany (2007).
[9] M. M. de Lima, Jr. and P. V. Santos, Rep. Prog. Phys. **68**, 1639 (2005).
[10] M. M. de Lima, Jr., W. Seidel, H. Kostial, and P. V. Santos, J. Appl. Phys. **96**, 3494 (2004).

6. Acknowledgments

This work was supported by the grant from German BMBF (grant No. 13N9891).

IOP Conf. Series: Materials Science and Engineering **41** (2012) 012010 doi:10.1088/1757-899X/41/1/012010

Influence of the Surface Pretreatment on the Electrical Properties of MgO and Al_2O_3 Gate Stacks grown by MBE

Chen-Yi Su, M Menghini, T Smets, L Dillemans, R Lieten and J P Locquet[1]

[1] Department of Physics and Astronomy, Katholieke Universiteit Leuven, Celestijnenlaan 200D, B-3001, Leuven, Belgium

chenyi.su@gmail.com

Abstract. For CMOS devices, the interface between the dielectric and the semiconductor is of crucial importance for the electrical properties. This paper is devoted to examining the role of different substrate preparation procedures on the interface properties. After the surface treatment 5 nm amorphous oxide films were deposited by molecular beam epitaxy (MBE). MgO on Si was chosen as our main model system. Additionally, a higher κ dielectric, Al_2O_3, was also studied. The electrical properties were investigated through measurements of MOS capacitor structures. Mainly capacitance density measurements as a function of voltage C-V were performed and analysed. Moreover, post-metal annealing experiments were performed in order to change the interface configuration and to optimize the electrical properties. The results demonstrate that the interface properties can be well controlled by the pretreatment of the substrates. Surprisingly it was also found that an identical surface pretreatment – despite the different dielectrics – induced a systematic, reproducible and substantial V_{FB} shift of the C-V curves while most other parameters remain unaffected. This opens additional possibilities to fine-tune the flat-band voltage in such high-κ transistors.

Introduction

With the continuing need to scale down the dimensions of CMOS devices while keeping high device performance, the requirements on the dielectrics such as a high capacitance and low leakage is getting more challenging [1]. A novel material with high dielectric constant and high band gap is required. Many high-κ materials have been studied for gate stack application, such as HfO_2, Al_2O_3, etc. However, in most high-κ systems, it is found that a sub silicon oxide layer is still necessary at the interface with silicon. Since the SiO_x layer acts as a series capacitor, the overall capacitance is lowered. Moreover, in MOS devices, different metal gates lead to different effective work function (EFW). Proper interface dipoles or charges must then be introduced to fine-tune the overall flat-band voltage [2]. In this research report, we aim to control the oxide semiconductor interfaces to improve and tune the electrical properties of MOS capacitors.

In order to atomically control the growing surface, an MBE system is used for the thin film deposition. We grew the oxide films using three different initial monolayers – oxygen rich, metal rich, and co-deposition. Those are indicated as O_2 ML, Mg ML, and MgO ML for the MgO system; O_2 ML, Al ML, and Al_2O_3 ML for the Al_2O_3 system respectively.

Experiment

B doped p-type Si (100) wafers with dopant concentration of 1×10^{15} cm^{-3} were used as substrates. The deposition of thin MgO films on 2 × 1 reconstructed Si (100) wafers was reported in detail elsewhere [3]. After 2 minutes of cleaning in a 0.5 % HF solution, the Si wafers were heated up to 570°C for 5 minutes in the MBE system for degassing. The deposition of MgO films was done from an Mg containing effusion cell at a substrate temperature of 350 °C by using molecular oxygen at a chamber pressure of 5×10^{-5} mbar and with a deposition rate of 0.1 Å / s. The three different initial monolayers mentioned above were deposited at the initial stage for each sample. In the case of the O_2 ML the sample surface was exposed first to O_2 for 1 minute, while in the case of the Mg ML one ML of Mg was deposited on the substrate before the oxygen was introduced. In the case of the MgO ML both O_2 and Mg were deposited on the substrate surface at the same time. The same growth conditions were also used for Al_2O_3 films using Al from an Al containing effusion cell.

The as grown films were then annealed at different temperatures (250 °C, 350 °C, and 450 °C) under forming gas (5 % H_2 and 95 % N_2) or N_2 gas for 15 minutes in order to study the annealing effect on the electrical properties. For electrical characterization, 40 nm Pt was deposited by MBE through a shadow mask to create the top electrodes. The backside of the samples was scratched and the samples were then glued on a metal plate with an eutectic glue for back contact. Top electrodes of 40 × 40 μm^2 area were used for capacitance-voltage (C-V) measurements by a Keithley 4200 system.

Results

With such a high oxygen pressure, amorphous films were grown and investigated by *in-situ* Reflection High Energy Electron Diffraction (RHEED) and *ex-situ* X-Ray Diffraction (XRD) [3]. When the films have a moderate leakage current, the C-V and interface-trap density, D_{it} can be studied systematically. Figure 1 (a) shows the C-V curves of the as grown MgO films on p-type Si substrates. A negative V_{FB} shift with higher oxygen content in the first monolayers was found. This phenomenon is even more pronounced for Al_2O_3 films as shown in Figure 1 (b). The flat-band voltage of MgO films are shifted but still very close to the theoretical V_{FB}, whereas there is a relatively large positive shift for the Al_2O_3 system. C-V curves after different annealing conditions are compared in Figure 2. The flat-band voltage gradually shifts to negative direction with increase annealing temperature. Note that at 450 °C annealing in N_2 causes almost the same effects as annealing in forming gas (FGA) on the V_{FB} shift. The same tendency versus annealing was found for samples with the O_2 ML, Mg ML, and MgO ML as the initial monolayer. The V_{FB} difference among all the initial growing conditions almost disappears after 450 °C annealing, and all samples have a V_{FB} close to 0 V compared with the theoretical value of 0.6 V (not shown here). As shown in Figure 2 (b), the annealing effects in Al_2O_3 system are similar as in the MgO system. The flat-band voltages of O_2 ML and Al_2O_3 ML films shift back to the theoretical value after 15 min 450 °C FGA (not shown here).

(a) (b)

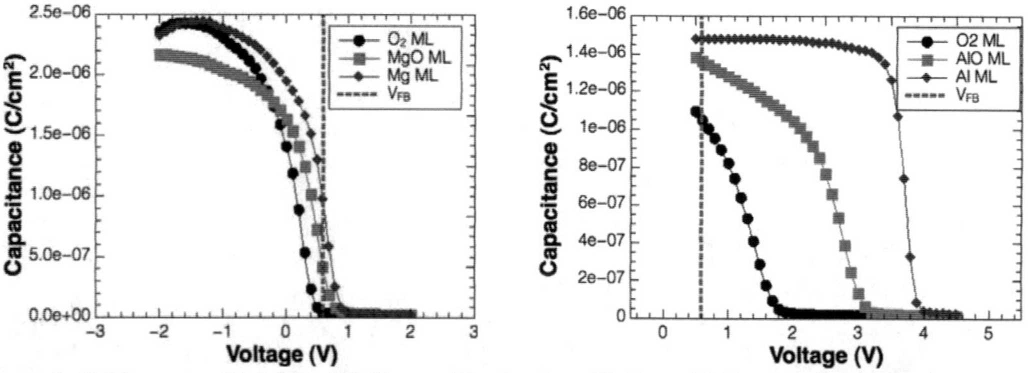

Figure 1. C–V curves of (a) 5 nm MgO on p-Si substrate; (b) 5 nm Al_2O_3 on p-Si substrate.

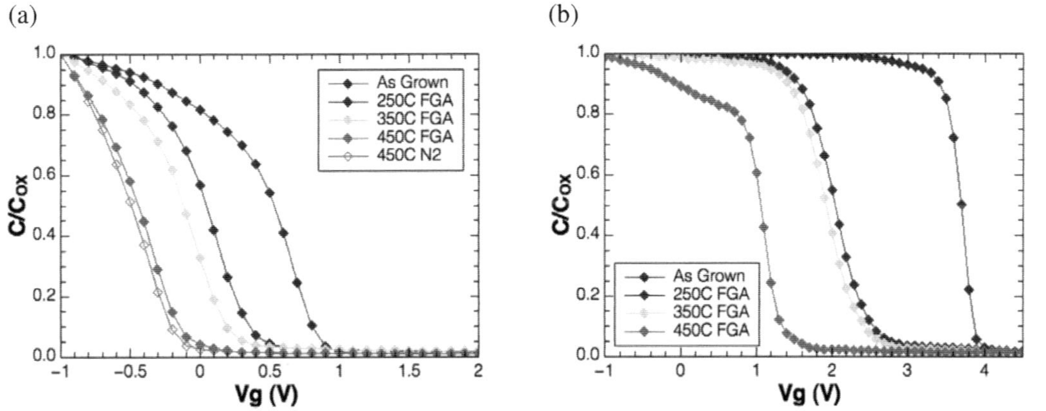

Figure 2. C-V curves after different annealing conditions (a) Mg ML MgO films; (b) Al ML Al_2O_3 films.

Discussion

Several models have been developed to explain the V_{FB} shift. One of them is the oxygen vacancy model [4]. When a lower valence atom replaces another atom with a higher valence, a positively charged oxygen vacancy is created according to the formula $MgO = Mg_{Si}'' + O_O^x + V_O^{\bullet\bullet}$. These positive charges stay close to the interface between Si substrate and the oxide layer, thus causing a negative shift of the flat-band voltage.

It has been found that a thin metal oxide capping layer can passivate the Si surface from further oxidation [5]. A metal rich initial monolayer prevents the Si surface from further oxidation much more effectively than an oxygen rich initial monolayer. As a result, thicker SiO_x films are formed under an oxygen rich initial growing environment. When lower valence metal atoms, ie. Mg or Al, diffuse into the SiO_x layers, thicker SiO_x films lead to more positive oxygen vacancies, and thus induce a larger negative V_{FB} shifts. Therefore O_2 ML films have the largest negative shifts. These processes are shown schematically in Figure 3. This phenomenon is more pronounced after annealing creating more vacancies, which thus causes larger negative shifts. Notice that diffusion is mainly a thermal process; this explains that the same effects are caused by FGA and N_2 annealing at the same temperatures.

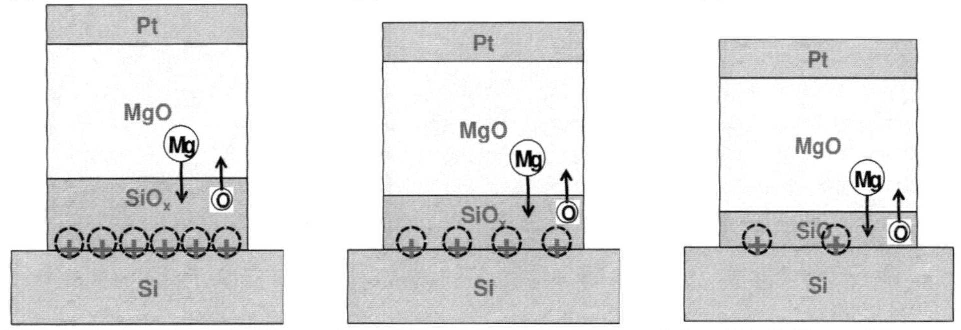

Figure 3. A schematic illustration of the oxygen vacancy model on MgO films with different initial condition (a) O_2 ML; (b) MgO ML; (c) Mg ML.

However, the oxygen vacancy model can not explain the initial systematically positive shift of V_{FB} in the Al_2O_3 system. To explain this behavior we consider another model, which is called the oxygen density model [6]. For two different oxides in contact with each other, there is a tendency to move oxygen atoms from a high oxygen density layer to a low oxide density layer. A positively charged vacancy is created. Together with the negative oxygen atom, a dipole is then formed. The amount and polarization of the dipoles defines then the direction and the amount of the V_{FB} shifts. Since both Al_2O_3 and MgO have higher oxygen density than SiO_2, oxygen atoms move towards SiO_x in both cases, as seen in Fig. 4. The effect leads to negative charges at the interface, and thus causes positive flat-band voltage shifts. Because Al_2O_3 has a higher oxygen density than MgO, more dipoles are formed in the Al_2O_3 system than in the MgO system, thus a more pronounced positive V_{FB} shift is found. The atomistic model proposed by L. Lin and J. Robertson [7] can also explain the opposite V_{FB} shift direction of Al_2O_3 and MgO. In this model, the shifts correlate with the work function / electronegativity of the metal of the metal oxide due to an electric field built by image charges. However, these latter two models cannot be used to explain the more negative V_{FB} shift of the O_2 ML compared to the Mg ML as discussed before. Clearly the observed results can only be explained by considering that the two different mechanisms are operational at the same time.

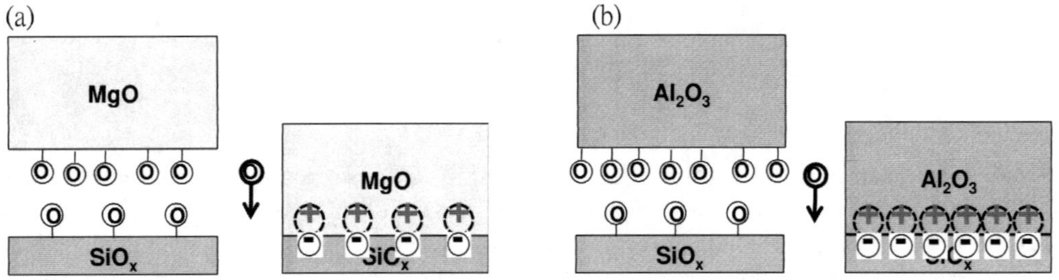

Figure 4. The Oxygen density model on (a) MgO on SiO_x; (b) Al_2O_3 on SiO_x.

Conclusion

In this research, three different initial surface treatments are performed, those are a metal ML first, a metal oxide ML first, and an oxygen ML first. We found that for both MgO and Al_2O_3 on Si, more negative V_{FB} shifts happen when more O_2 is introduced in the initial growth. This can be explained by positive oxygen vacancies formed at the interface. V_{FB} roll-off (negative shift) happens after a post deposition annealing, indicating more thermal diffusion and that more oxygen vacancies are formed. The systematically positive V_{FB} shift in the Al_2O_3 system is due to its relatively high oxygen density and / or electronegativity. In conclusion, with a suitable initial surface treatment and annealing step, it is possible to control the flat-band voltage of MOS capacitors.

References

[1] Locquet J P, Marchiori C, Sousa M, Fompeyrine J and Seo J W 2006 *J. Appl. Phys.* **100** 051610
[2] Narayanan V *et al* 2006 *VLSI Symposium Tech. Dig.* **224**
[3] Su C Y, Frederickx M, Menghini M, Dillemans L, Lieten R, Smets T, Seo J W and Locquet J P 2012 *Thin Solid Films* **520** 4508-11
[4] Guha S, Paruchuri V K, Copel M, Narayanan V, Wang Y Y, Batson P E, Bojarczuk N A, Linder B and Doris B 2007 *Appl. Phys. Lett.* **90** 092902
[5] Marchiori C, Frank M M, Bruley J, Narayanan V and Fompeyrine J 2011 *Appl. Phys. Lett.* **98** 052908
[6] Kita K and Toriumi A 2009 *Appl. Phys. Lett.* **94** 132902
[7] Lin L and Robertson J 2009 *Appl. Phys. Lett.* **95** 012906

Semiconducting YBaCuO films grown on silicon substrates: IR room temperature sensing and fast pyroelectric response

A J Kreisler[1,3], V S Jagtap[1,4], G Sou[2], G Klisnick[2] and A F Dégardin[1]
[1] SUPELEC / LGEP; CNRS UMR 8507; UPMC Univ Paris 06; Univ Paris Sud 11; 11 rue Joliot-Curie, Gif sur Yvette, 91190, France
[2] L2E; UPMC Univ Paris 06; 65-66 E2, 4 place Jussieu, 75252 Paris Cedex 05, France

E-mail: alain.kreisler@supelec.fr

Abstract. $YBa_2Cu_3O_{6+x}$ (YBCO) oxides, well known as superconductors for $x > 0.5$, are semiconductors (SC) for lower oxygen content. SC films DC sputtered at 150 °C in amorphous form (a-YBCO) exhibited a temperature coefficient of resistance competitive with other bolometric sensing materials. Such deposition conditions are highly beneficial to integrate a radiation detector on a silicon chip bearing already processed readout electronics. IR detector structures have been processed: simple planar or more advanced metal/a-YBCO/metal trilayers. IR response was investigated at 850 nm as a function of the modulation frequency. The planar structure exhibited a regular bolometric response below 100 Hz, then a pyroelectric high-pass response up to 100 kHz. The trilayer device only exhibited the pyroelectric behavior up to 100 kHz. In the pyroelectric regime, detectivity values peaked above 2×10^8 cm $Hz^{½}$/W with time constants in the μs range, under both DC biased and unbiased conditions.

1. Introduction

There has been a constant interest, during the last decade, in room temperature (RT) semiconducting IR bolometric detectors. Being sensitive to power, they are in principle wavelength independent, an advantage for wide spectral operation, whereas photon detectors only operate beyond the absorption cutoff. Among the popular semiconductors, we find amorphous silicon a-Si [1] and vanadium oxide families [2]. In practice, co-integration of the sensing device (or device array) with silicon readout electronics is also a key feature that would *a priori* give advantage to a-Si (or a-SiGe [3]) sensors because of the low deposition temperature of the amorphous material, compatible with already processed readout circuitry. Y-Ba-Cu-O oxide (in its oxygen depleted, non superconducting form i.e. $YBa_2Cu_3O_{6+x}$ with x below 0.3) has also raised interest: it offers competitive sensitivity and noise level and can be easily elaborated in amorphous form (a-YBCO) at 100-200 °C on silicon based substrates [4,5]. Moreover, a-YBCO exhibits pyroelectric properties [4], which paves the way to an alternate thermal sensing detector type, with advantages we wish to stress in the present paper.

2. a-YBCO film preparation and material properties

a-YBCO films were DC sputtered using a hollow Y-Ba-Cu-O stoichiometric cathode under a Ar + O_2 atmosphere (total pressure: 33 to 67 Pa, Ar:O_2 flow ratio 55%:45%). The growth rate increased from

[3] To whom any correspondence should be addressed.
[4] Present address : Lab. MPQ, Univ. Paris Diderot, 10 rue A Domont & L Duquet, 75205 Paris Cedex 13, France

60 nm/h at 67 Pa to 160 nm/h at 33 Pa, with no noticeable change in the film quality [6]. Insofar as the bolometric device application is concerned, the temperature coefficient of resistance $TCR = (1/R)(dR/dT)$ is a key parameter to monitor the quality of the films. *TCR* as a function of deposition temperature is shown in figure 1. Substrates were either MgO (001) single crystal or p-doped silicon (1-10 Ω·cm resistivity) bearing a thermal SiO_x (500 nm) layer.

It appears that deposition temperatures in the 100-200 °C lead to amorphous films of low roughness (3 to 5 nm rms) exhibiting 20 nm grains immersed in a seemingly homogenous matrix [7].

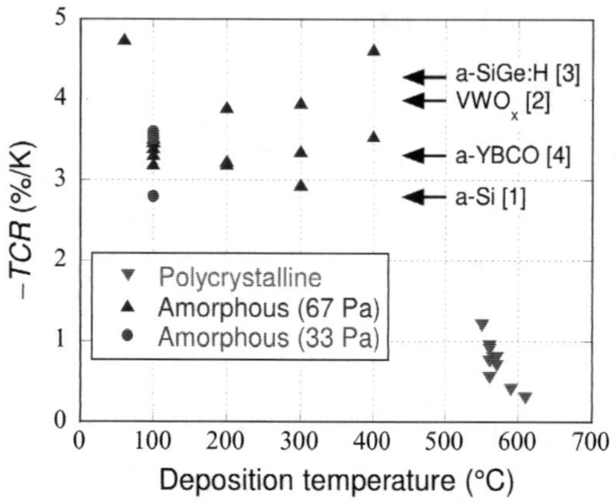

Figure 1. For DC sputtered Y-Ba-Cu-O films, temperature coefficient of resistance (*TCR*) as a function of deposition temperature T_d. For T_d above 550 °C, the films were polycrystalline and exhibited a low *TCR*. Whereas for $T_d < 500$ °C, the films were amorphous with good *TCR* values. For $100 < T_d < 200$ °C, no noticeable effect of substrate nature or film thickness was observed. The arrows indicate other published data. After [6].

3. Device geometries and electrical characteristics

In order to test a-YBCO IR sensing performance, two device geometries were considered; both were elaborated on p-Si/SiO_x substrates, as depicted in figures 2 and 3.

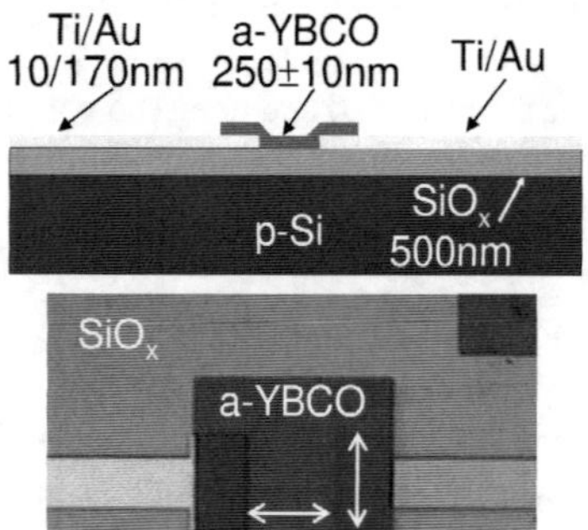

Figure 2. Cut-view (top) and photograph (bottom) of the planar device (DevP).

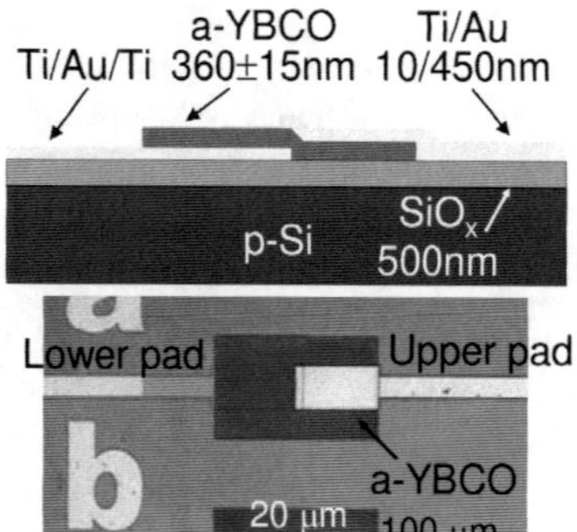

Figure 3. Cut-view (top) and photograph (bottom) of the trilayer device (DevT).

In the planar configuration (DevP), a-YBCO was in-plane connected to two gold pads (figure 2). The DC *I*(*V*) characteristic exhibited a close to symmetric linear variation for $|V| > 2$ V, with a resistance slope of 12- 13 MΩ and a resulting resistivity $\rho_{DC} \cong 300$ Ω·cm. The low voltage non-linearity was fitted with a $I \propto V^{1.75}$ power law due to a bulk effect (not to a contact effect, see below) related to a-YBCO structural disorder, the film behaving as an assembly of resistive elements, randomly connected e.g. at grain boundaries. Such a model provides 1 to 2 exponent values [8].

A trilayer configuration (DevT) was designed to decrease the device access resistance (figure 3). a-YBCO was sandwiched between gold contacts with an extra titanium layer. The DC *I*(*V*) characteristic was highly non-linear, we ascribed to the Schottky nature of the contacts. At low level, a very good fit $I \propto \exp(V^{1/4})$ was obtained for both polarities, which well models a reverse biased Schottky diode with image force barrier lowering [9]. The device actually presents a reverse biased junction for both polarities (back to back connected Schottky diodes). Assuming the p-type nature of a-YBCO, as the work function for Ti (4.3 eV) is lower than for Au (5.1 eV), the metal/a-YBCO junction would be Schottky for Ti (DevT) but ohmic for Au (DevP).

From $\rho_{DC}(T)$ variations (T = 100-300 K) in the (quasi-)linear regime (V_{DC} = 8 V for DevP and 3.7 V for DevT), an Arrhenius behavior was observed as $\rho_{DC} \propto \exp(E_A/k_BT)$. From the activation energy we deduced $TCR = -E_A/(k_BT^2) = -2.8$ %/K at 300 K for DevP and -2.0 %/K for DevT. The degradation with respect to bare films is significant for DevT, due to the more critical processing sequence.

4. Near-IR characterization results and discussion

The optical response was measured at 850 nm, with modulated power at frequency f. The device was voltage biased at V_{DC} and the delivered current i_S was readout with a transconductance preamplifier, which output was detected at reference frequency f with a lock-in amplifier (also used to measure the device 1 Hz bandwidth noise current density i_N). Other details can be found elsewhere [5,10].

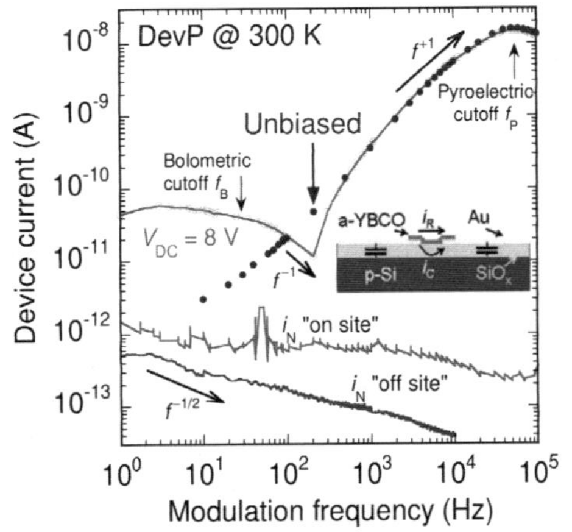

Figure 4. NIR response of DevP (biased and unbiased) as a function of modulation frequency. "Off site" refers to measurement in an electromagnetically quiet location. Inset: the substrate behaves as a counter-electrode for the capacitance current path.

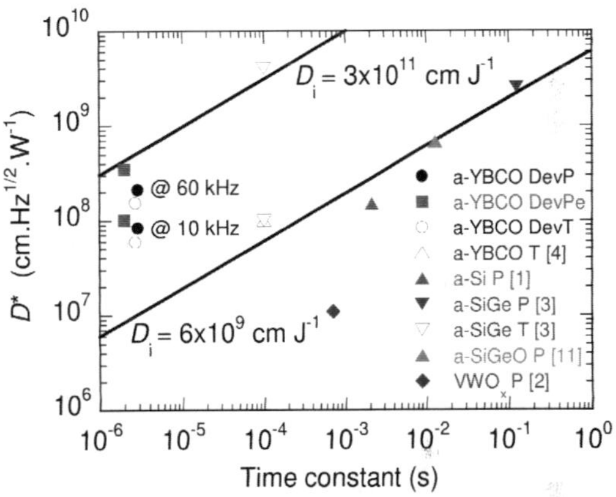

Figure 5. For the various devices listed (P: planar, T: trilayer), detectivity as a function of time constant. The first three symbols refer to the present study (enhanced conditions – DevPe – are detailed in [10]). D_i represents the impulse detectivity, as commented in the text.

Results for DevP are shown in figure 4. For V_{DC} = 8 V bias, the regular bolometric response (thermal variation of resistance) exhibits the expected low-pass behavior (cutoff $f_B \cong 30$ Hz). As DevP is not a suspended structure, the bolometric response is low. A sharp rise of the response is observed above 300 Hz, with a high-pass behavior (cutoff $f_P \cong 55$ kHz) in line with a-YBCO pyroelectricity [4]. The thermal variation of the electric polarization could be detected because a capacitance current i_C is collected due to the Si-substrate acting as a (floating) counter-electrode (figure 4, inset). In fact, with an insulating MgO substrate, only the bolometric response was observed [5]. Moreover, from unbiasing the device (V_{DC} = 0) resulted a pure pyroelectric response without signal degradation.

For DevT, only the pyroelectric response was observed due to the intrinsic capacitor structure of this device (figure 3), the resistive (regular bolometric) current becoming negligible ($i_R << i_C$).

The device detectivity was obtained from the current responsivity i_S/P_I (P_I is the effective incident IR power) and i_N (on-site, figure 4) as $D^* = A_D^{1/2} i_S/(i_N P_I)$ (A_D is the effective device area). Our pyroelectric response data are collected in figure 5 with other published results [1-4,11]. Plotting D^* vs. detector time constant τ allows to illustrate the impulse detectivity parameter $D_i = D^* \tau^{-1/2}$ (in $cm \cdot J^{-1}$) that is an invariant for a given class of detectors (same working principle, same background noise temperature [12]). $D_i = 6\times10^9$ $cm \cdot J^{-1}$ represents an average for a majority of published results that mainly concern suspended devices optimized for low frequency operation [1,3,4,11]. Our data are gathered around more than one order of magnitude higher, due to the fast pyroelectric response.

Among other points that would deserve addressing in detail, let us cite: i) the origin of the short response time (fast heat removal might be related to i_C flowing in the Si substrate and/or to the metal contacts); ii) the relationship between the pyroelectric cutoff frequency and the upper dielectric relaxation frequency observed at a few 10 kHz [13]; iii) the correlation between the dipole charges (origin of pyroelecticity) and the charge carriers photo-induced above the optical gaps observed in [7].

5. Conclusion

Semiconducting a-YBCO is an attractive sensing material for IR thermal detection. a-YBCO thin films can be deposited on silicon substrates and thus allow integration with CMOS readout electronics. Planar and tri-layer structures have been processed and tested in the near IR. Only the planar structure exhibited a regular bolometric response at low frequency. However, both structures exhibited a high-pass pyroelectric behavior, with detectivity values above 2×10^8 $cm \cdot \sqrt{Hz}/W$ and 2-3 μs time constants – whether DC biasing the device or not. Besides, many issues concerning the detecting principles (with respect to a-YBCO material properties and device geometry) still remain open.

The pyroelectric high frequency sensitivity offers a promising solution for fast imaging applications, especially in the far IR / THz range. Improvements are under consideration: for front-end radiation coupling with THz micro-antennas [14], the advantage of the trilayer resides in its lower access impedance. For back-end electric coupling, they deal with high impedance/low noise readout circuitry and correction of the lower frequency response distortion as well.

Acknowledgments

This work has been supported in part by a Marie Curie EST Fellowship of the European Community's FP-6 under contract number MEST-CT-2005-020692. It has benefited from a specific help of UPMC Univ Paris 06. The devices have been realized at the CTU-Minerve facilities, University Paris Sud 11 (Orsay, France). Mr P. Thoreau, student at UPMC, is thanked for experimental help.

References

[1] Liu X-M, Fang H-J and Liu L-T 2007 *Microelectronics Journal* **38** 735-739
[2] Chi-An N and Moon S 2007 *Infrared Phys. Technol.* **50** 38-41
[3] Moreno M, Kosarev A, Torres A and Ambrosio R 2007 *Thin Solid Films* **515** 7607-7610
[4] Butler D P, Çelik-Butler Z, Jahanzeb A, Gray J E and Travers C M 1998 *J. Appl. Phys.* **84** 1680-1687
[5] Jagtap V S, Longhin M, Kulsreshath M K, Kreisler A J and Dégardin A F 2010 *Terahertz Physics, Devices and Systems IV: Proc. of SPIE* vol 7671 ed M Anwar *et al* 76710X
[6] Longhin M, Kreisler A J and Dégardin A F 2008 *Materials Science Forum* **587-588** 273-277
[7] Jagtap V S, Dégardin A F and Kreisler A J 2012 *Thin Solid Films* **520** 4754-4757
[8] Roux S and Herrmann H J 1987 *Europhys. Lett.* **4** 1227-1231
[9] Rhoderick E H 1980 *Metal-semiconductor contacts* (Oxford: Clarendon Press) pp 86-88
[10] Kreisler A J, Jagtap V S and Dégardin A F 2012 *Physics Procedia* to be published
[11] Ahmed A and Tait R N 2005 *Infrared Phys. Technol.* **46** 468-472
[12] Khrebtov I A 1991 *Sov. J. Opt. Technol.* **58** 261
[13] Gensbittel A, Dubrunfaut O, Jagtap V J, Kreisler A J and Dégardin A F 2012 *Thin Solid Films* **520** 4749-4753
[14] Scheuring A, Türer I, Ribière-Tharaud N, Dégardin A F and Kreisler A J 2010 *Ultra-Wideband Short Pulse Electromagnetics 9* ed F Sabath *et al* (New-York: Springer) pp 277-286

Epitaxial PZT thin films on YSZ-buffered Si (001) substrates for piezoelectric MEMS or NEMS applications

Corentin Jorel[1], Héloïse Colder[2], Alice Galdi[1,3,4], Laurence Méchin[1]

[1] GREYC, ENSICAEN, UCBN, CNRS, UMR 6072, 6 bd du Maréchal Juin, 14050 Caen cedex, France

[2] CRISMAT/CNRT-Matériaux /ENSICAEN, UMR6508, UMS 3318, 6 bd du Maréchal Juin, 14050 Caen cedex, France

[3] CNR SPIN Superconductors Oxides & Other Innovat M, I-84084 Fisciano, SA, Italy

[4] Univ Salerno, I-84084 Fisciano, SA, Italy

E-mail: corentin.jorel@unicaen.fr

Abstract. We report the growth of epitaxial $Pb(Zr_{0.54}Ti_{0.46})O_3$ (PZT) thin films on yttria-stabilized zirconia buffered silicon substrates by pulsed laser deposition. We demonstrate a full in plane epitaxy of the buffer layer, showing a RMS roughness of less than 0.3 nm for a 120 nm thick layer. This buffer layer allows the growth of fully (110) textured oxide conducting $SrRuO_3$ and subsequent functional oxide layers. Here the $Pb(Zr,Ti)O_3$ oxide was chosen to demonstrate its possible integration in piezoelectric microelectromechanical systems on silicon.

1. Introduction

High quality functional oxides are usually deposited on specific ceramic substrates such as $SrTiO_3$ (STO), $LaAlO_3$, MgO, etc., which are expensive and not compatible with silicon-based semiconductor technologies or suitable for high frequency applications. Considerable efforts have been made to develop deposition processes of epitaxial functional oxides on Silicon substrate.

Yttria-stabilized zirconia (YSZ) can be used as a buffer layer for the epitaxial growth of oxides on silicon substrates, such as $YBa_2Cu_3O_{7-\delta}$ (YBCO) [1, 2], $La_{0.7}Sr_{0.3}MnO_3$ (LSMO) [3,4], $Pb(Zr,Ti)O_3$ (PZT) [5]. Here we present results obtained by using YSZ buffer layers on silicon for improved piezoelectric PZT thin films by pulsed laser deposition (PLD). To avoid dielectric constant reduction, serious fatigue degradation and blocking contact as commonly observed when using noble metal (Pt) electrodes [6], we considered the $SrRuO_3$ (SRO) oxide as epitaxial conducting electrodes because of its lattice and chemistry match with PZT.

2. The YSZ buffer and SRO conductive electrode layers

All layers were deposited in situ by PLD on silicon (001) substrates in a multitarget system without breaking the vacuum using a pulsed KrF excimer laser (λ = 248 nm). YSZ crystallises in a 5.139 Å cubic lattice. SRO is a pseudocubic perovskite with a pseudocubic lattice parameter of 3.93 Å (orthorhombic with *Pbnm* symmetry and lattice parameters a=5.567Å, b=5.5304Å and c=7.8446Å).

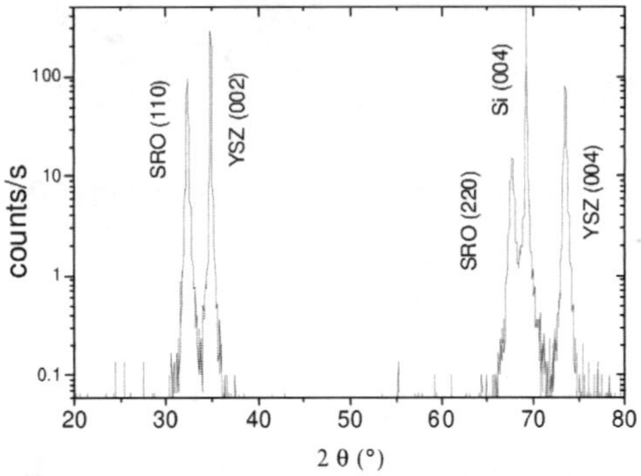

Figure 1. X-ray diffraction patterns in the θ–2θ configuration of SRO / YSZ layers on a silicon (001) substrate.

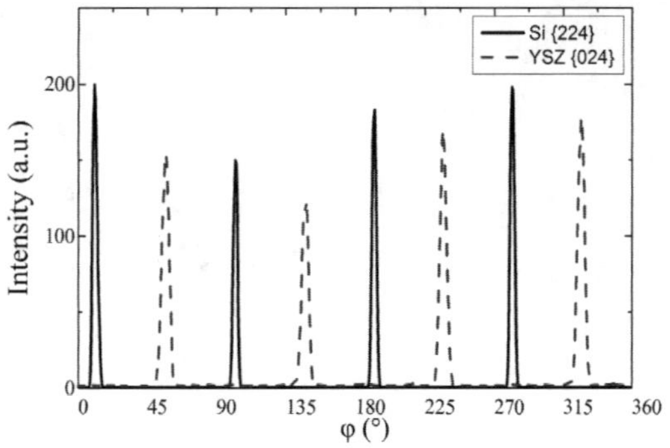

Figure 2. φ-scan diffraction patterns of the YSZ {024} and Si {224} families of plane.

Prior to deposition, the 10 × 10 mm^2 Si substrates were ultrasonically degreased in acetone and alcohol without any removal of the native oxide. They were mounted on a heater block and heated radiantly. In order to avoid a further growth of the SiO_2, the Si substrates were heated up to the deposition temperature at a residual pressure of 10^{-5} mbar. The pressure was increased to 10^{-4} mbar after 30 seconds of deposition of YSZ, which enabled to remove most of the native SiO_2 layer by a chemical reaction with ZrO_2 or Zr in vacuum during the first laser shots [7]. The epitaxy could then be established in spite of the primitive SiO_x amorphous layer. The YSZ deposition temperature was fixed to 700°C, the laser pulses and energy at 3 Hz and 220 mJ, respectively. In these optimized deposition conditions, (001) oriented YSZ films were obtained (figure 1). They showed rocking curves full width at half maxima (FWHM) of the YSZ (002) peak of 0.6°.

Figure 3. 5 × 5 μm^2 AFM images in tapping mode of a 120 nm thick YSZ layer on Si (001). RMS roughness: 0.291 nm.

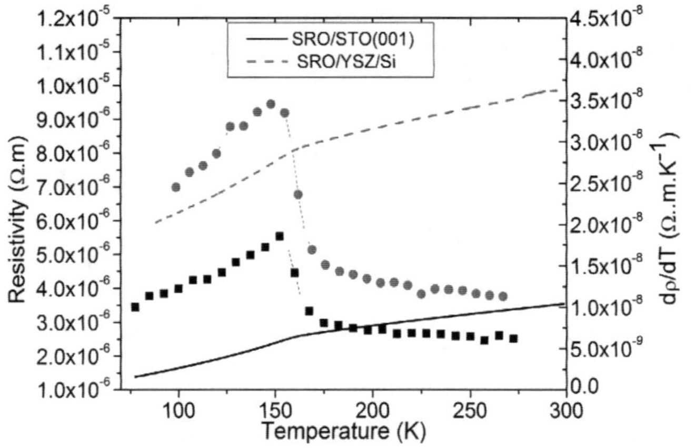

Figure 4. Resistivity versus temperature characteristics (lines) of 30 nm thick SRO films deposited on YSZ/Si and on STO (001) substrates as a reference. The derivatives are plotted on the right axis (scatter).

The in-plane epitaxial relationships of the YSZ films have been investigated by φ-scan XRD patterns (figure 2). We have considered the {224} families of planes for silicon, and the {024} for

YSZ. The φ-scan XRD patterns were recorded with the rotation angle φ parallel to the [001] direction of each material. As shown in Fig. 2, the shifting by 45° between the YSZ {024} and Si {224} families of planes indicates the overlap of all the directions with the same *(hkl)* indices (YSZ [100]//Si [100] for example). YSZ layers therefore showed a full in-plane epitaxy with a cube on cube alignment with respect to the Si substrate. Finally, YSZ buffer layers showed a low rms roughness of 0.291 nm, calculated with the AFM data presented figure 3.

The perovskite structure, the compatible lattice parameter, the high conductivity and the chemical compatibility of SRO made it a great candidate for conducting electrodes [8]. SRO was deposited by PLD at 700 °C, 2 Hz and 170 mJ in an oxygen pressure of 0.2 mbar. Typical layers are 20-80 nm thick for 1000-3000 laser pulses.

XRD patterns of the SRO/YSZ bilayer on Si (001) are shown in figure 1 and demonstrate a high degree of crystallinity of the (110) single orientation. Corresponding rocking curves give 0.7° mean dispersion of the growth axis (FWHM) for the SRO film, which is close to the value of 0.6° obtained for the YSZ film.

The electrical behavior of the SRO electrode was checked through the resistivity vs. temperature four point measurements shown in figure 4. As a comparison the characterization of an equivalent film deposited on STO (001) is also shown. Both films show metal-like behavior and resistivity values at ambient temperature are state of the art [8]: 360 and 1000 μΩ.cm for the film grown on STO and Si respectively.

3. PZT films

$PZ_{0.54}T_{0.46}$ films were grown on the SRO/YSZ bilayer at 600°C, 5Hz, with 200 mJ laser pulses in different O_2 atmospheres ranging from 0.05 to 0.5 mbar. Due to Pb loss during the PLD process [9], a 10% PbO enriched target was used. The a and c lattice parameters of $PZ_{0.52}T_{0.48}$ are 4.036 and 4.146 Å respectively so that the lattice mismatch with SRO is fairly small (~2.7 %).

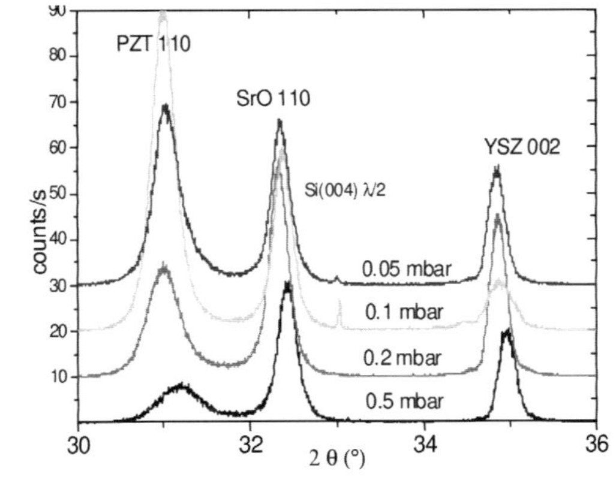

Figure 5. X-ray diffraction patterns in the θ–2θ configuration of PZT/SRO/YSZ layers on Si for different oxygen pressure.

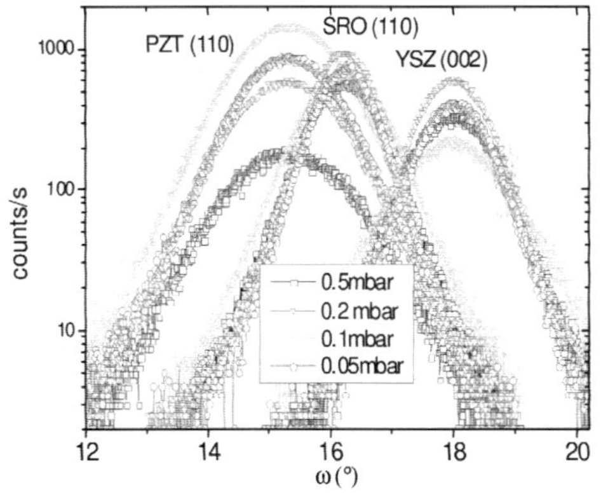

Figure 6. Corresponding rocking curves

XRD diffraction results are presented figures 5 and 6. All films were fully (110) oriented, growing accordingly with the SRO electrode layer. Best results were obtained at the lower deposition pressure (0.05 mbar) as shown in table 1 below. Moreover, the less crystallized 0.5 mbar film displays parasitic micrograins on AFM scans. The low dispersion of results indicates that PZT are robust to deposition conditions, which is favorable for future applications.

Table 1. Rocking curve FWHM of the films at different oxygen deposition pressures

P(mbar)	YSZ	SRO	PZT
0.05	1.2	1	1.7
0.1	1.8	1.4	1.8
0.2	1.2	1	1.8
0.5	1.4	1.2	2.1

4. Conclusion & perspectives

Epitaxial SRO/YSZ bilayers were grown by PLD on silicon (001) substrates. Both films type showed excellent crystalline properties and SRO exhibited a fully metal-like behavior.

This bilayer constitutes a very versatile building block for further functional oxide deposition. We showed that fully oriented (110) PZT could be grown on top of it. Further work has now to be conducted for the use of these layers in microelectromechanical and nanoelectromechanical devices [10].

References

[1] L. Méchin, J.C. Villégier, G. Rolland, F. Laugier, *Physica C* **269** 124-30 (1996)
[2] L. Méchin, G. Huot and D. Bloyet, *Appl. Phys. Lett.* **85** 3154 (2004)
[3] L. Méchin, P. Perna, C. Barone, J.-M. Routoure, Ch. Simon, *Mater. Sci. And Engine. B* **144** 73-7 (2007)
[4] P. Perna, L. Méchin, M.P. Chauvat, P. Ruterana, Ch. Simon, and U. Scotti di Uccio, *J. Phys. : Condens. Mat.* **21** 306005 (2009)
[5] M. Dekkers, M.D. Nguyen, R. Steenwelle, P.M. te Riele, D.H.A Blank, and G. Rijnders, *Appl. Phys. Lett.* **95** 012902 (2009)
[6] R. Ramesh, W.K. Chan, B.Wilkens, H. Gilchrist, T. Sands, J.M. Tarascon, V.G. Keramidas, D.K. Fork, J. Lee, and A. Safari, *Appl. Phys. Lett.* **61** 1537 (1991)
K. Yokota, H. Morigou and F. Miyashita, *Nucl. Ins. & Method. In Phys. Res. B* **257** 468 (2007).
L. Qiao and X. Bi, *J. of All. & Comp.* **477** 560 (2009).
B. Chen, H. Yang, J. Miao, L. Zhao, L. X. Cao, B. Xu, X. G. Qiu, and B. R. Zhao, J. *Appl. Phys.* **97** 024106 (2005).
[7] S. J. Wang, C. K. Ong, L. P. You, and S. Y. Xu, *Semicond. Sci. Technol.* **15** 836 (2000).
[8] G. Koster, L. Klein, W. Siemons, G. Rijnders, J.S. Dodge, C.-B. Eom, D.H.A. Blank, M.R. Beasley, *Rev. Mod. Phys.* **84** 253–98 (2012)
[9] R. Thomas, S. Mochizuki, T. Mihara, T. Ishida, *Mater. Sci. And Engine. B* **95** 36 (2002)
I.M. Reaney, K. Brooks, R. Klissurska, C. Pawlaczyk, N. Setter, *J. Am. Ceram. Soc.* **77** 1209 (1994)
[10] C. Ayela, T. Alava, D. Lagrange, D. Remiens, C. Soyer, T. Ondarcuhu, A.Greve, L.Nicu, *IEEE Sensors J.* **8** (3) 210 (2008)
T Alava, F Mathieu, L Mazenq, C Soyer, D Remiens and L Nicu, *J. Micromech. Microeng.* **20** 075014 (2010)

Broadband dielectric and IR pyroelectric response of amorphous Y-Ba-Cu-O oxygen depleted thin films

A Gensbittel[1], A F Dégardin[1], O Dubrunfaut[1], M K Kulsreshath[1], V S Jagtap[1] and A J Kreisler[1]

[1] SUPELEC; CNRS UMR 8507; UPMC Univ Paris 06; Univ Paris Sud 11; LGEP, 11 rue Joliot-Curie, 91190 Gif-sur-Yvette, France

E-mail: Alain.Kreisler@supelec.fr

Abstract. We report on the study of dielectric properties of amorphous semiconducting $YBa_2Cu_3O_{6+x}$ ($x < 0.5$, YBCO) thin films with two different thicknesses measured in a broad frequency range (from 40 Hz to 2 GHz) and in the 210 to 430 K temperature range, using a coaxial discontinuity technique. At all temperatures, dielectric permittivity and conductivity spectra exhibit typical features of dielectric relaxation processes. A first relaxation at low frequency (300 Hz – 2 kHz), observed only at high temperature, might be attributed to interfacial effects. Grain boundaries of the YBCO thin films could explain the second relaxation observed at higher frequency (800 Hz – 660 kHz). The higher relaxation frequencies increase with temperature and film thickness and follow a thermally activated behavior of the Arrhenius-type. Application to IR detection with a film deposited on silicon has been targeted: the observed pyroelectric behavior of the detector exhibits a fast response in the μs range.

1. Introduction

Understanding how the disorder structure in amorphous semiconductors influences their electrical, dielectric and optical properties is of great concern to develop high performance electronic and photonic devices. The semiconducting phase of $Y_1Ba_2Cu_3O_{6+x}$ (YBCO) with a low oxygen content ($x < 0.5$) has shown promising performances as an uncooled infrared (IR) bolometric detector [1] but also as a fast pyroelectric detector [2]. The aim of the present work is to study dielectric properties of amorphous YBCO (hereafter referenced to as *a*-YBCO) films in the 40 Hz to 2 GHz frequency range and in the 210 to 430 K temperature range, using a coaxial discontinuity technique. Results are discussed with respect to the film microstructure and direct-current (DC) electrical behavior. The pyroelectric response of an *a*-YBCO device in the near-IR (NIR) range is also investigated.

2. Experimental procedure

a-YBCO thin films were deposited at room temperature by off-axis DC sputtering using a hollow cathode technique [3]. 500 nm thick (001) MgO substrates and 380 μm thick *p*-doped silicon substrates (coated with a thermally grown 500 nm thick SiO_x layer) were used.

The film surface morphology was observed by atomic force microscopy in the tapping mode (Digital Instruments AFM Nanoscope® III). Dielectric characterization in reflection mode was performed using a cylindrical closed cell technology, described in detail in [4]. For this purpose, YBCO films were deposited on MgO circular substrates of 7 mm diameter. The side of the sample facing and abutting the inner conductor of an APC-7 coaxial connector was coated with a gold circular pad. A bare MgO substrate was prior measured to subtract its influence. Dielectric measurements were conducted with an HP 4294A impedance analyzer below 110 MHz and an Agilent PNA 8364B network analyzer above 45 MHz. Temperature was varied from 210 K to 430 K. For optical characterization, a 100 μm × 100 μm sensing area planar structure was patterned in a 250 nm thick *a*-YBCO film grown on a *p*-Si/SiO_x substrate. The optical response was measured at 850 nm wavelength with a VCSEL diode, which output power was modulated electronically. The device current was readout with a low input resistance trans-conductance amplifier (SR 570) the output voltage of which was synchronously detected at reference frequency *f* with a lock-in amplifier (SR 830 up to 100 kHz and SR 844 above 100 kHz).

3. Results and discussion

3.1. Structural and DC electrical transport properties

As shown in figure 1, AFM imaging revealed the granular texture of the amorphous films. By looking closely, we notice that grains were isolated in a seemingly uniform amorphous matrix. The average grain size d_{AFM} increased as the film thickness t_{film} increased, thus leading to films with a lower roughness r_{film}: d_{AFM} = 29 nm±5 nm and $r_{film} \approx 9$ nm rms for t_{film} = 500 nm; d_{AFM} = 59 nm±5 nm and $r_{film} \approx 3$ nm rms for t_{film} = 880 nm. The typical DC resistivity value at 300 K was 350 Ω·cm [5] and its temperature behavior was of Arrhenius-type with an activation energy E_a value of 0.26 eV for the 500 nm thick film and 0.24 eV for the 880 nm thick film.

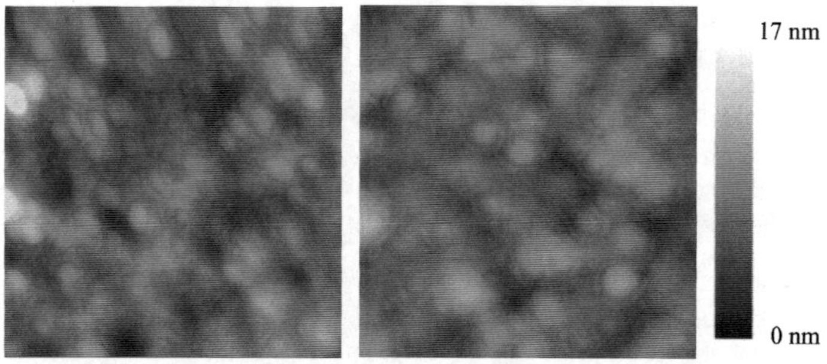

Figure 1. AFM images (0.25×0.25 μm^2 area) of 500 nm-thick (left) and 880 nm-thick (right) a-YBCO films.

3.2. Dielectric properties

The real part ε' and imaginary part ε'' of the complex relative permittivity of the YBCO films were extracted from admittance or reflection coefficient measurements. As displayed in figure 2a, ε' exhibited typical features of dielectric relaxation processes for both films: ε' started decreasing gradually with increasing frequency *f* up to 20 kHz (200 kHz, respectively) at temperature T = 300 K (430 K, respectively); then ε' sharply dropped until reaching a stable value of ≈ 1000 above 200 kHz (6 MHz, respectively) at 300 K (430 K, respectively). By looking closely at the frequency independent trend of ε' below 200 Hz at 430 K, we can suspect, however, the presence of another relaxation mechanism. The latter was only visible above 330 K in our measurements, in line with low frequency data published in [6]. Moreover, ε' values were on the whole larger for the 500 nm thick film.

The evolution of the loss tangent $\tan \delta = \varepsilon'' / \varepsilon'$ as a function of the frequency is displayed in figure 2b for selected temperatures. Measurements evidenced distinct peaks related to a dielectric loss process, which shifted to a higher frequency when temperature increased (≈ 200 kHz at 300 K, ≈ 6 MHz at 430 K) in line with the evolution of ε' with *f*.

To extract relaxation frequency values, data were fitted with a Debye model and represented on Cole-Cole plots, as described in detail elsewhere [7]. According to this model, the overall relative permittivity around a dielectric relaxation can be written as $\varepsilon_r = \varepsilon_L + \varepsilon_H$, where ε_L is the low frequency permittivity before the fall-off and ε_H the high frequency permittivity after the fall-off. Figure 3 shows one example of Cole-Cole plot (i.e. $\varepsilon''(f)$ versus $\varepsilon'(f)$) obtained for the 880 nm thick film at 360 K. Two relaxation frequencies are evidenced: f_{rel1} = 669 Hz leading to $\varepsilon_{H1} = 9\times10^4$. The second one is observed at f_{rel2} = 129 kHz leading to $\varepsilon_{H2} \approx 800$. Between these relaxations, the value $\varepsilon_{L2} = 8\times10^4$ is very close to ε_{H1}. Figure 4 shows the evolution of these relaxation frequencies f_{rel} as a function of temperature. In line with our observation in $\varepsilon'(f)$ plot, f_{rel} which increased with temperature followed a thermally activated behavior of the Arrhenius type as $f_{rel} = f_0 \exp(-E_a/k_BT)$, where f_0 is a frequency pre-factor and k_B the Boltzmann constant. For the 500 nm (880 nm, respectively) thick film, E_a and f_0 values obtained were E_{a1} = 0.25 eV (0.24 eV, respectively) and f_{01} = 4 MHz (1.5 MHz, respectively) from the fit on the first relaxation frequency f_{rel1}. They were E_{a2} = 0.29 eV (0.26 eV, respectively) and f_{02} = 900 MHz (820 MHz, respectively) from the fit on the second relaxation frequency f_{rel2}.

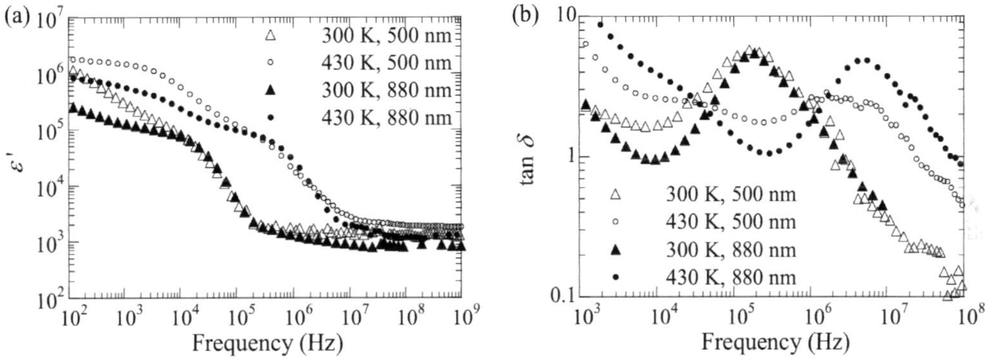

Figure 2. Dispersion of (a) the real part ε' of the complex dielectric permittivity and (b) loss tangent $\tan\delta$ in *a*-YBCO films displayed here at selected temperatures.

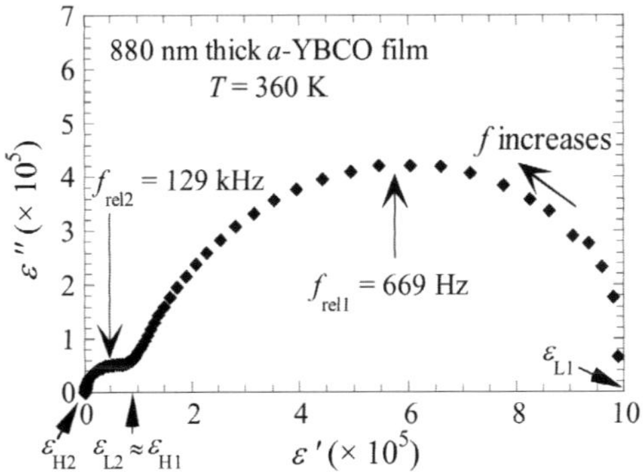

Figure 3. Cole-Cole plot (imaginary part ε'' vs. real part ε') of the complex permittivity.

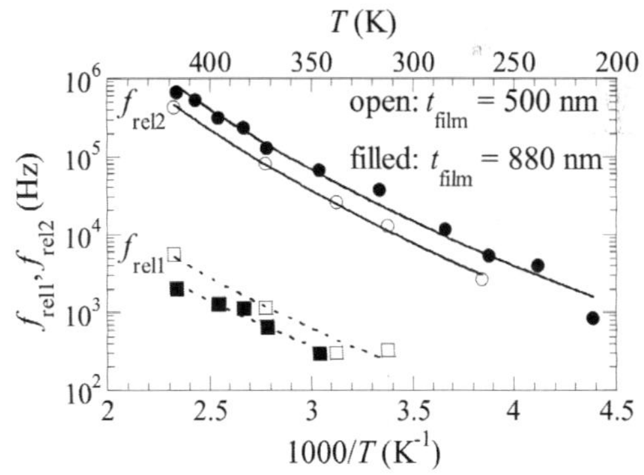

Figure 4. Arrhenius plots of the relaxation frequencies (Debye model).

The similar E_a values, as extracted from Arrhenius-type evolution of the DC resistivity on one hand and relaxation frequency f_{rel1} and f_{rel2} on the other hand, comfort the description of the sources of polarization in *a*-YBCO films by a hopping process of localized holes [8] assisted by phonons of vibrational frequency f_0, according to the Pike model [9]. The close to three orders of magnitude difference between the f_{01} and f_{02} values suggest, however, distinct hopping mechanisms characterized by largely different hopping ranges. Whereas the slow (f_{rel1}) relaxation process might be attributed to interfacial effects at electrical contacts, the faster (f_{rel2}) one could be assigned to grain (or sub-grain) boundaries acting as localized impurity sites, as expected for amorphous semiconductors.

IOP Conf. Series: Materials Science and Engineering **41** (2012) 012013 doi:10.1088/1757-899X/41/1/012013

The difference between both films, although deposited under the same conditions on MgO substrates, might be interpreted for the slow (f_{rel1}) relaxation process by a variation in the roughness of the gold contact pad, such a variation of the gold / *a*-YBCO interface inducing a shift of the relaxation frequency. The difference between both films for the faster (f_{rel2}) relaxation process might originate from the film microstructure, which revealed an increasing average grain size and a smoother surface when increasing film thickness, thus leading to a decrease of the barrier height (hence a lower activation energy value) of this thermally activated relaxation mechanism.

3.3. NIR pyroelectric behavior

As shown in figure 5, the signal current measured at 300 K on an unbiased *a*-YBCO planar device vs. modulation frequency exhibited a purely high-pass behavior ($\propto f^1$ increase followed by f_P = 55 kHz cut-off / τ_P = 2.9 μs) that could be assigned to the pyroelectric state of *a*-YBCO. This value is to be compared with the frequency values of the fast relaxation process (f_{rel2} = 12-37 kHz at 300 K) which was attributed to YBCO grains, so reflecting the intrinsic character of pyroelectricity in *a*-YBCO.

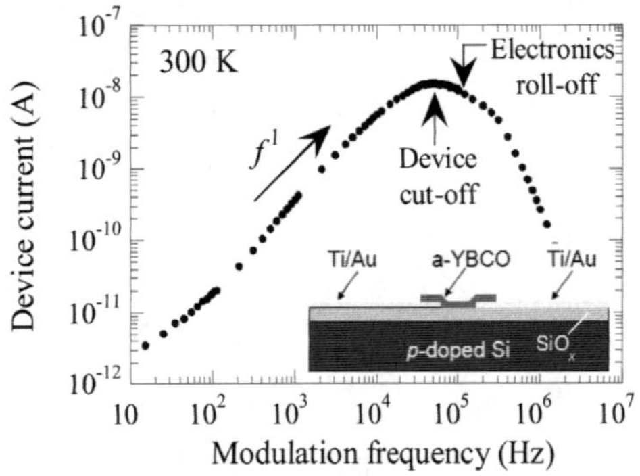

Figure 5. NIR response (signal current) of a planar device (made from a 250 nm thick *a*-YBCO film) measured at room temperature as a function of modulation frequency. The device was unbiased i.e. directly connected to the current preamplifier input.

4. Conclusion

Dielectric characterizations of semiconducting *a*-YBCO thin films have evidenced two dielectric relaxations: a first one at low frequency might be attributed to effects at the YBCO / gold contact interface; a second one observed at higher frequency could be due to the granular structure of the amorphous films. The NIR response of an *a*-YBCO planar device exhibited a high-pass pyroelectric response behavior with a 2-3 μs time constant, which could offer a promising solution for fast imaging applications, especially in the far IR / THz range where moderate cost systems should be considered.

5. Acknowledgments

This work has been partially supported by a Marie Curie EST Fellowship of the EC FP-6 under contract number MEST-CT-2005-020692. It has benefited from a specific help of UPMC Univ Paris 06. The device has been realized at the CTU-Minerve facilities (Univ Paris Sud 11, Orsay, France).

6. References

[1] Kumar S, Chitteboyina M M and Butler D P 2007 *J Vac. Sci. Technol.* B **25** 1315-1320
[2] Kreisler A J, Jagtap V S and Dégardin A F 2012 *Physics Procedia* to be published
[3] Longhin M, Kreisler A J and Dégardin A F 2008 *Materials Science Forum* **587-588** 273-277
[4] Belhadj-Tahar N-E and Fourrier-Lamer A 1986 *IEEE Trans. MTT* **34** 346-350
[5] Jagtap V S, Dégardin A F and Kreisler A J 2012 *Thin Solid Films* **520** 4754-4757
[6] Gray J E, Çelik-Butler Z, Butler D P and Jahanzeb A 1998 *Ferroelectrics* **209** 517-539
[7] Gensbittel A, Dubrunfaut O, Jagtap V S, Kreisler A J and Dégardin A F 2012 *Thin Solid Films* **520** 4749-4753
[8] Samara G A, Hammetter W F, Venturini E L 1990 *Phys. Rev. B* **41** 8974-8980
[9] Pike G E 1972 *Phys. Rev. B* **6** 1572-1580

IOP Conf. Series: Materials Science and Engineering **41** (2012) 012014 doi:10.1088/1757-899X/41/1/012014

Synthesis and characterization of highly c-textured $Al_{(1-x)}Sc_{(x)}N$ thin films in view of telecom applications

M A Moreira[(1,2)], J Bjurström[(1)], V Yantchev[(1)], I Katardjiev[(1)]

[1] Dept. Solid State Electronics, Ångström Laboratory, Uppsala University, Box 534, SE-751-21, Uppsala, Sweden
[2] Center for Semiconductor Components (CCS), University of Campinas, P.O. Box 6061, João Pandiá Calógeras 90, 13.083-870, Campinas-SP, Brazil

E-mail: milena.moreira@angstrom.uu.se

Abstract. Wurtzite AlN is a piezoelectric material with excellent electro-acoustic properties and is used for the fabrication of high frequency thin film micro-acoustic components, most notably filters, duplexers, resonators, etc. Its moderate electromechanical coupling coefficient (k_t^2) of 6%-7% is insufficient for applications requiring larger bandwidths. Recent theoretical and experimental studies indicate that AlN alloyed with Sc exhibits a substantially higher piezoelectric constant than pure AlN. This study aims at determining the main electro-acoustic parameters of $Al_{(1-x)}Sc_{(x)}N$ in view of large bandwidth applications. To this end, highly c-textured $Al_{(1-x)}Sc_{(x)}N$ thin films have been synthesized with relative Sc concentrations of up to 0,15. Subsequently, FBAR resonators were fabricated and characterized as a function of the Sc content. It is seen that k_t^2 increases linearly with the latter to a value of 12% for a Sc concentration of x=0,15, while the Q value decreases from 739 to about 348 in the same concentration range. Likewise, the TCF varies from -35,9ppm/°C to -39,8ppm/°C, while the dielectric constant increases from ε=10 to a value of 14,1 for x=0,15. Finally, the relative dielectric losses are seen to increase by approximately a factor of two.

1. Introduction

Wurtzite thin film aluminum nitride (w-AlN) is a well-known material used in commercial frequency control components due to its excellent electroacoustic properties such as moderate piezoelectric constant e_{33} of 1.55 C/m^2 yielding an electromechanical coupling coefficient k_t^2 of 6.5%, low dielectric and acoustic losses (Q > 1000) and high acoustic wave velocity. There is a growing number of applications requiring larger bandwidths resulting in a sustained interest in materials with higher piezoelectric constants while still exhibiting low overall losses.

Recent studies have demonstrated both theoretically and experimentally an increase in the piezoelectric response of AlN films alloyed with scandium (Sc) [1][2]. These studies were followed by first-principle calculations [3] and subsequent experimental measurements of the dielectric permittivity [4] which indicated an expected increase in the electro-mechanical coupling (k_t^2) as a direct consequence from an increased piezoelectric constant (e_{33}) in conjunction with a decreased stiffness constant (C_{33}). These expectations were confirmed experimentally [6] by direct measurements of the electro-mechanical coupling in a thin film bulk acoustic resonator (FBAR) configuration.

The goal of this work is to provide further details regarding the thin film growth reported in [6] as well as perform further characterization of films properties in terms of temperature coefficient of frequency and dielectric losses.

2. Experimental

One port thin film bulk acoustic resonators (FBAR) were fabricated on 100 mm in diameter, 300 µm thick double side polished (100) silicon wafers. A 100 nm thick c-textured AlN seed layer followed by a 300 µm thick molybdenum (Mo) film were sputter deposited onto the RCA cleaned silicon substrates without external heating. The Mo film is not patterned and serves as a bottom (ground) electrode to the FBAR structures. Deposition of around 1 µm thick $Al_{(1-x)}Sc_{(x)}N$ films was performed in a dual magnetron, high vacuum system (< $2x10^{-8}$ Torr) at 500 °C. Two metallic targets (Al 99.99%, Ø 100 mm and Sc 99.9%, Ø 100 mm) were reactively co-sputtered using pulsed DC mode in an argon (Ar 30 sccm) and nitrogen (N_2 30 sccm) gas mixture at a constant process pressure of 2.8 mTorr. The Al and Sc atomic concentration of the resulting film were controlled by adjusting the discharge power on the respective target.

Contact via holes through the $Al_{(1-x)}Sc_{(x)}N$ films down to the bottom Mo electrode were opened by wet etching in TMAH at 50 °C using a 500 nm thick sputtered silicon dioxide as a hard mask. Thick aluminum metal contact pads (~1 µm thick) and top electrode (~200 nm thick) were then sputter deposited and patterned to complete the resonator. The final process step of releasing the resonator membrane is anisotropic dry etch of the substrate from the backside using an Al hard mask. A note on the sideline regarding the seed AlN layer in the above process sequence is in order. Its purpose is threefold. Firstly, it serves as a buffer layer to prevent inter-diffusion of Mo and Si in subsequent process steps at elevated temperatures. Secondly, it acts as an efficient etch stop during the silicon substrate backside etch. Thirdly, it has been shown, that a thin AlN seed layer improves the degree of texture of the Mo film, which in turn results in an improved crystallographic texture of the active AlN film subsequently deposited onto the latter [1].

Four films with a thickness of 2µm with four different $Al_{(1-x)}Sc_{(x)}N$ film compositions (x = 0; 0.03; 0.09 and 0.15) were grown. The composition was determined by X-ray Photoelectron Spectroscopy (XPS) analysis while crystallographic properties were determined from θ-2θ scans and rocking curve measurements using a Philips XPert Pro MRD X-ray diffractometer (XRD).

The beneficial effect of the AlN seed layer on the texture of the $Al_{(1-x)}Sc_{(x)}N$ films is seen from Figure 1a,b showing the full width at half maximum (FWHM) of the (002) rocking curve for x = 0.09 and x = 0.15 respectively for both with and without AlN seed layer.

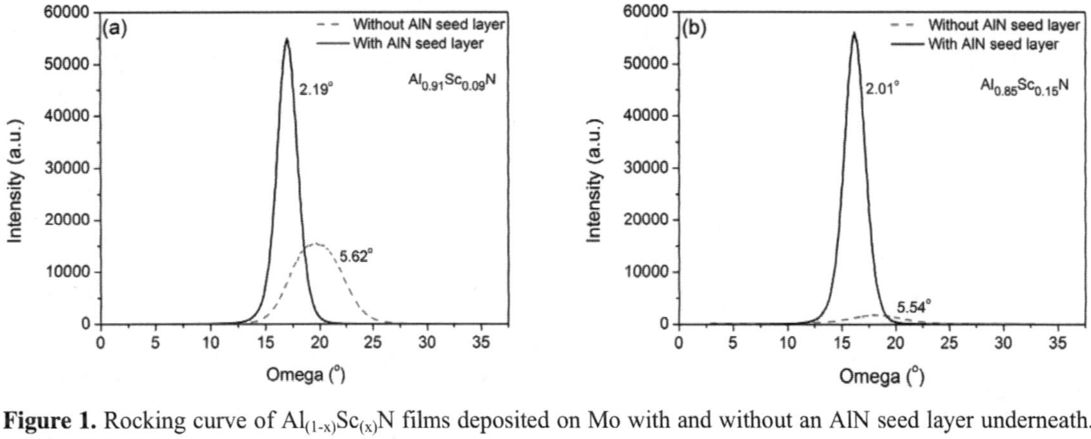

Figure 1. Rocking curve of $Al_{(1-x)}Sc_{(x)}N$ films deposited on Mo with and without an AlN seed layer underneath. (a) x = 0.09; (b) x = 0.15.

The θ-2θ scans show that the films are c-axis oriented for all compositions (see fFigure 2. The FWHM value of the (002) rocking curve was in the range of 1.6° for low Sc concentration films up to 2° for $Al_{0.85}Sc_{0.15}N$. Additional diffraction peaks are identified as the planes from the metal electrodes (Mo and Al) and from the Si(100) substrate.

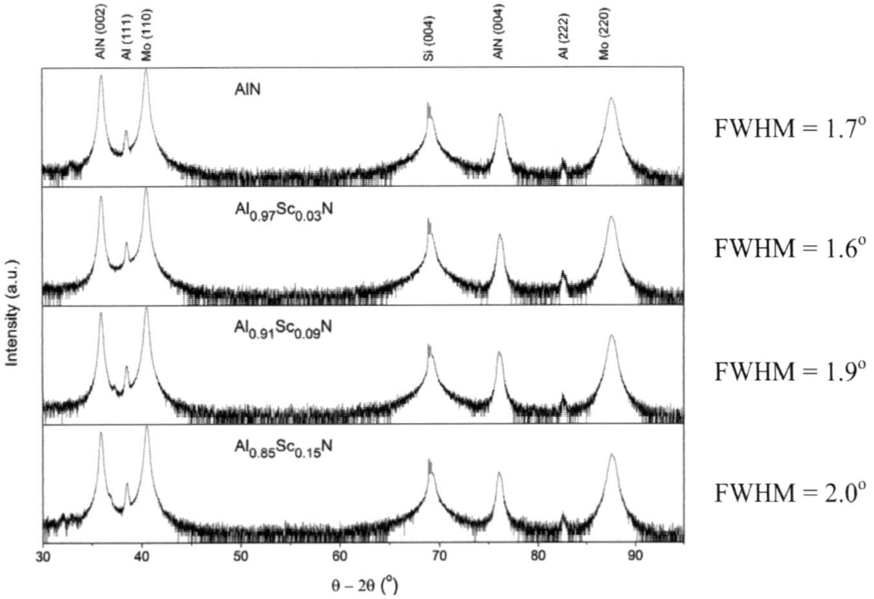

Figure 2. θ-2θ scans of $Al_{(1-x)}Sc_{(x)}N$ films with different Sc concentrations (x = 0 up to x = 0.15).

3. Electroacoustic characterization

The electro-acoustic characterization of the fabricated FBARs was carried out with an Agilent E8364B network analyzer equipped with a 150 μm pitch pico-probe and calibrated for a 50 Ω system. One port reflection (S11) measurement data were acquired at frequencies around the fundamental thickness excited longitudinal mode. Figure 3. FBAR impedance response vs. frequency for different compositions of $Al_{(1-x)}Sc_{(x)}N$ films shows the measured FBAR impedance response versus frequency for various compositions of the piezoelectric film.

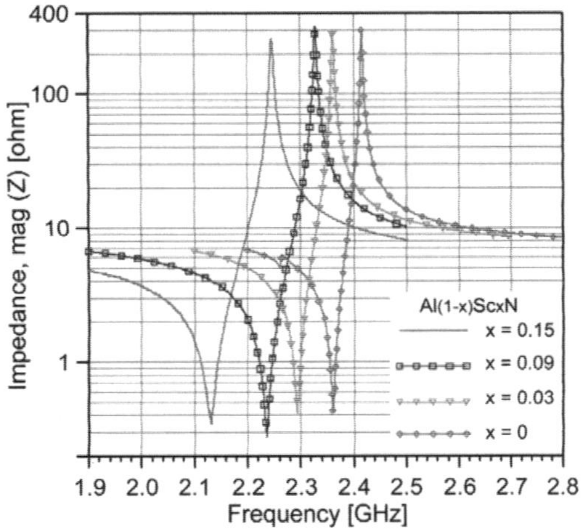

Figure 3. FBAR impedance response vs. frequency for different compositions of $Al_{(1-x)}Sc_{(x)}N$ films.

The above measurements have been fitted in appropriate models [6] to extract important material constants such as piezoelectric constant e_{33}, stiffness constant C_{33}, dielectric permittivity ε_r as well as to characterize device performance in terms of Q value and electro-mechanical coupling k_t^2 as illustrated in Table 1. Measured FBAR quantities (f_s, TCF, tan δ) and extracted film parameters (ε_r, C_{33}, e_{33}, Q, k_t^2) for various compositions of the $Al_{(1-x)}Sc_{(x)}N$ films.

It is seen that the Q value decreases with increasing Sc concentration. Noteworthy, the device figure-of-merit, FOM = k_t^2.Q, is found to exhibit a maximum FOM at a relative Sc concentration of x = 0.09 which illustrates the potential of this material for high performance devices.

Table 1. Measured FBAR quantities (f_s, TCF, tan δ) and extracted film parameters (ε_r, C_{33}, e_{33}, Q, k_t^2) for various compositions of the $Al_{(1-x)}Sc_{(x)}N$ films.

Composition	f_s (GHz)	ε_r	C_{33} ($x10^{11}$ N/m^2)	e_{33} (C/m^2)	Q	k_t^2 (%)	TCF (ppm/°C)	tan δ ($x10^{-3}$)
AlN	2.3605	10.17	3.70	1.48	739	6.16	-35.9	1.82
$Al_{0.97}Sc_{0.03}N$	2.2935	10.95	3.46	1.66	601	7.55	-35.2	1.65
$Al_{0.91}Sc_{0.09}N$	2.2339	12.37	3.26	1.94	513	9.53	-37.1	2.78
$Al_{0.85}Sc_{0.15}N$	2.1522	14.12	2.70	2.14	348	12.00	-39.8	4.96

The temperature coefficient of frequency (TCF) was measured in the interval 25-125 °C and is seen to increase somewhat with Sc concentration.

Finally, metal-insulator-metal (MIM) capacitors were fabricated using $Al_{(1-x)}Sc_{(x)}N$ films in the same concentration interval for measurement of the dielectric losses (tan δ) as also shown in Table 1. Measured FBAR quantities (f_s, TCF, tan δ) and extracted film parameters (ε_r, C_{33}, e_{33}, Q, k_t^2) for various compositions of the $Al_{(1-x)}Sc_{(x)}N$ films.

4. Conclusions

$Al_{(1-x)}Sc_{(x)}N$ FBARs and MIM capacitors with four different Sc concentrations, namely x=0;0.03; 0.09 and 0.15 respectively have been fabricated and electrically characterized. It is found that both the electro-mechanical coefficient and the dielectric losses increase by a factor of 2 with increasing Sc concentration while the device Q value decreases correspondingly in the concentration range studied. The TCF of $Al_{(1-x)}Sc_{(x)}N$ films is seen to increase slightly with Sc concentration.

References

[1] Akiyama M, Kamohara T, Kano K, Teshigahara A, Takeuchi Y and Kawahara N 2009 Enhancement of piezoelectric response in scandium aluminum nitride alloy thin films prepared dual reactive cosputtering *Adv. Mater.* **21** 593-96

[2] Akiyama M, Kano K and Teshigahara T 2009 Influence of growth temperature and scandium concentration on piezoelectric response of scandium aluminum nitride alloy thin films *Appl. Phys. Lett.* **95** 162107-1-3

[3] Tasnadi F, Alling B, Höglund C, Wingqvist G, Birch J, Hultman L and Abrikosov I A 2010 Origin of the anomalous piezoelectric response in wurtzite $Sc_xAl_{1-x}N$ alloys *Phys. Rev. Lett.* **104** 137601-1-4

[4] Wingqvist G, Tasnadi F, Zukauskaite A, Birch J, Arwin H and Hultman L 2010 Increased electromechanical coupling in w-$Sc_xAl_{1-x}N$ *Appl. Phys. Lett.* **97** 112902-1-3

[5] Kamohara T, Akiyama M, Ueno N and Kuwano N 2008 Improvement in crystal orientation of AlN thin films prepared on Mo electrodes using AlN interlayers *Ceram. Int.* **34** 985-9

[6] Moreira M, Bjurström J, Katardjiev I and Yantchev V 2011 Aluminum scandium nitride thin-film bulk acoustic resonator for wide band applications *Vaccum* **86** 23-6

IOP Conf. Series: Materials Science and Engineering **41** (2012) 012015 doi:10.1088/1757-899X/41/1/012015

Metal-Insulator-Metal capacitors with ALD grown $SrTiO_3$: Influence of Pt electrodes

M. Lukosius[a], T. Blomberg[b], D. Walczyk[a], G. Ruhl[c], Ch. Wenger[a]

[a] IHP, Im Technologiepark 25, 15236 Frankfurt (Oder), Germany

[b] ASM Microchemistry Ltd., Väinö Auerin katu 12A, 00560 Helsinki, Finland

[c] Infineon Technologies AG, Wernerwerkstr. 2, 93049 Regensburg, Germany

email: lukosius@ihp-microelectronics.com

Abstract. Metal-Insulator-metal (MIM) capacitors with atomic layer deposited (ALD) single $SrTiO_3$ dielectric have been investigated in this work. Structural and electrical properties were studied after the formation of the MIM stack consisting of the platinum (Pt) bottom electrode, 50 nm $SrTiO_3$ layer and the top Pt electrode. The as deposited films were amorphous and had a dielectric constant of ~ 10, whereas the annealing of the samples in the nitrogen (N_2) or oxygen (O_2) atmosphere at 550-600 °C led to the crystallization of the $SrTiO_3$ and therefore to the increased dielectric constant of ~ 85. In addition, the electrical results revealed that the combination of $SrTiO_3$ with the high work function electrode like Pt, provided better leakage current performance in comparison with TiN/ $SrTiO_3$ stacks. The values as low as ~ 10^{-7} A/cm^2 at 2 V were observed for both in N_2 or O_2 annealed $SrTiO_3$ layers. On the other hand, the samples annealed in O_2 atmosphere at 600 °C possessed lower capacitance-voltage nonlinearity coefficients (-645 ppm/V^2) than the ones for N_2 annealed samples (-2700 ppm/V^2).

1. Introduction

As the progress and developments in wireless communication pushes the integration towards complete system-on-chip (SoC) units, Metal-Insulator-Metal (MIM) capacitors, which are one of the key passive elements in analog/mixed signal integrated circuits, have also to posses superior properties. According to the International Roadmap for Semiconductors (ITRS) for wireless communication technologies, the quadratic voltage-capacitance nonlinearity coefficient α must be smaller than 100 ppm/V^2, capacitance density should be higher than 5 fF/µm^2, quality factor more than 50 at 5 GHz as well as low leakage current density of 10^{-8} A/cm^2 [1]. These requirements imply the replacement of silicon oxide-based dielectrics with new high-k materials. A number of alternative materials, binary oxides like HfO_2 [2] or Al_2O_3 [3] and ternary systems like Sr-Ta-O [4] or Ti-Ta-O [5] have been studied for MIM applications, to name a few. Among the alternative dielectrics, $SrTiO_3$ is one of the most promising candidates due to its high dielectric constant of ~ 150 in the crystalline phase. Due to this high k value, the corresponding capacitance density of ~ 44 fF/µm^2 was observed for the MIM structures with 26 nm $SrTiO_3$ [6]. However, it is well known that grain boundaries in crystalline dielectric material create channels for electrons and therefore may lead to high leakage currents [7]. In addition, high leakage currents could be related to the small band gap of 3.5 eV of $SrTiO_3$ [8]. We have previously reported that slightly Sr rich, 50 nm $SrTiO_3$ thin films, grown by Atomic Layer Deposition (ALD) method, possessed a high dielectric constant of 95, but the leakage current density of TiN/$SrTiO_3$/Au stacks was 10^{-3} A/cm^2 at 1 V [9]. As this increased leakage current challenges the practical use of $SrTiO_3$ in MIM applications, in this work we have applied high work function metal bottom as well as top electrodes like platinum; in order to reduce the leakage currents of the $SrTiO_3$ based MIM capacitors. It is also known that electrical properties of MIM capacitors are not only dependent on the dielectric itself, but also on the bottom and the top electrodes. In addition, the effect of different post deposition anneals (performed in nitrogen or oxygen) on the structural and electrical properties of the $SrTiO_3$ layers, have been investigated in this study.

2. Experimental

Thin films of $SrTiO_3$ were deposited by ALD in an ASM Pulsar® 2000 type reactor at 250 °C using (1,2,4-tertiarybutyl-Cp)$_2$Sr and Ti(OMe)$_4$ metal precursors and O_3 as the oxidizing agent. Composition of the films was controlled by changing the cycle ratios of the SrO and TiO_2 depositions steps and the SrO/TiO_2 pulsing sequence of 2:1 was used in this study, which resulted in strontium content of ~60 at % Sr/(Sr+Ti). Films were deposited on ∅200 mm Pt/Ti/SiO_2/Si wafers with ~80 nm Pt and ~10 nm Ti layers.

The thicknesses of the investigated films were measured by means of ellipsometry. Typical thickness of $SrTiO_3$ was 50 nm in this study. After the depositions of $SrTiO_3$, the films were annealed in nitrogen or oxygen at 500 °C, 550 °C and 600 °C for 120 min in order to evaluate the crystallization behavior of the films. Film crystallinity was studied by X-ray diffraction (XRD). Finally, for electrical properties of metal-insulator-metal structures, Pt top electrodes with a diameter of 400 μm were sputtered at room temperature through a shadow mask. The capacitance-voltage (C-V) characteristics were measured at 100 kHz using Agilent 4294A precision impedance analyzer. The data for current-voltage (I-V) characteristics was collected at a probe station with a HP 4140 amperemeter.

3. Results

In the first step, the crystallization behavior of the $SrTiO_3$ layers was investigated by XRD after annealing the samples at 500 °C, 550 °C and 600 °C in nitrogen or oxygen atmospheres. As already expected, the amorphous character of as deposited samples was monitored and only Pt(111) and Si(200) reflections were visible as shown in Fig. 1 line a). After annealing of the samples at 500 °C in oxygen atmosphere (line b) in Fig. 1), broad (110) and (200) $SrTiO_3$ signals were visible, indicating some crystallization already at this temperature. Additional peak at ~ 27.5 ° was also observed. This signal was attributed to the rutile phase of TiO_2 indicating several possibilities of forming this TiO_2 phase. This could be related to the oxidation of the Ti adhesion layer under the platinum bottom electrode layer due to the long time of annealing or to the fact that the crystallization of $SrTiO_3$ started with TiO_2 formation. On the other hand we have previously investigated the crystallization behavior of $SrTiO_3$, deposited on TiN, where $SrTiO_3$ was completely crystalline, and no formation of TiO_2 was observed after the annealing of the samples at 600 °C for 5 minutes [9]. The increase of the annealing temperature to 550 °C in this study enabled the crystallization of $SrTiO_3$. Typical reflections, (100), (110), (200) and (211) of $SrTiO_3$ can be seen in the Fig. 1 line c). The XRD patterns of the samples, annealed at 600 °C were very similar to the ones annealed at 550 °C. At this point it has to be mentioned that the same results were also observed for the $SrTiO_3$ films, annealed in nitrogen atmosphere and the crystallization temperature of 550 °C was also observed. This temperature was also reported as a crystallization temperature for $SrTiO_3$ [10].

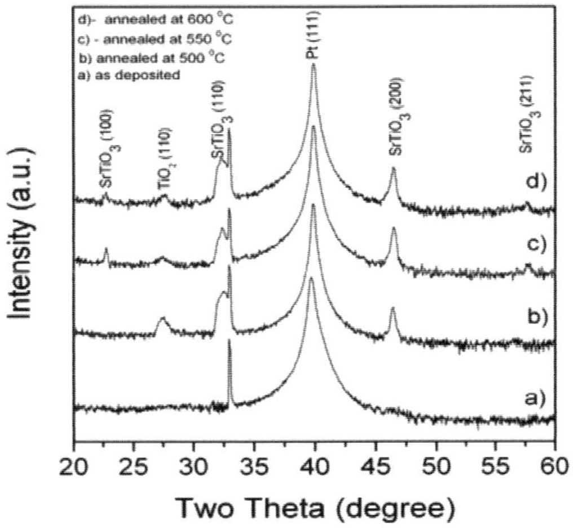

Fig. 1. XRD patterns of SrTiO3 films, deposited on Pt: a) as deposited at 250 °C, b) annealed at 500 °C, c) annealed at 550 °C and d) annealed at 600 °C

After the annealing of the samples, the top platinum electrodes were evaporated through the shadow mask to form MIM structures in order to extract electrical properties. C-V measurements were performed at 100 kHz in the range of -3 V to + 3V. Dielectric constant values, extracted from these measurements are plotted in Fig. 2 for as deposited and in oxygen (solid line in Fig. 2) or nitrogen (dotted line in Fig. 2) annealed layers. It can be seen that a permittivity values of ~ 10 are measured for the as deposited and at 500 °C annealed $SrTiO_3$ samples, whereas an abrupt increase of the k values to ~ 85 are observed if the annealing was performed at 550 °C. These results clearly correlate with the XRD measurements, where crystallization temperature of 550 °C for $SrTiO_3$ was also monitored. The dielectric constants slightly decreased for the samples annealed at 600 °C, indicating that long annealing time at this temperature worsened the electrical characteristics. In addition, Fig. 2 shows that there is only very small effect of the annealing atmosphere on the dielectric constant values of $SrTiO_3$ films. As the thickness of $SrTiO_3$ was 50 nm in this case, the corresponding capacitance density values of ~ 1.7 fF/μm^2 for as deposited and ~ 14 fF/μm^2 for crystalline $SrTiO_3$ were measured.

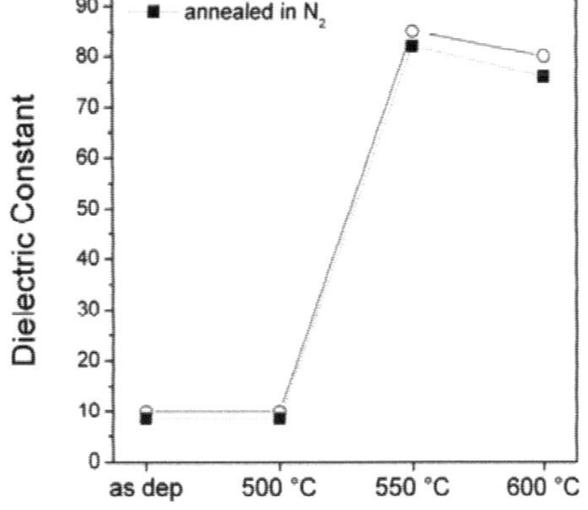

Fig. 2. Dielectric constant dependence on the annealing temperature for $SrTiO_3$ samples annealed in oxygen (o) and in nitrogen (■) atmosphere.

Leakage current is also very important parameter for MIM capacitors, especially at higher voltages. Its dependence on the annealing temperature and annealing atmosphere is presented in Fig. 3 a) and b). It is clearly seen that the lowest leakage current densities are measured for as deposited and at 500 °C annealed $SrTiO_3$ samples. The values as low as ~ 10^{-8}-10^{-9} A/cm^2 at ± 2V were observed for both in oxygen as well in nitrogen annealed $SrTiO_3$ based MIM capacitors. On the other hand, as already expected, leakage current densities increased by 2-3 orders of magnitude for the annealed samples, since crystalline layers typically possess higher leakage currents than amorphous ones. Moreover, the samples annealed at 550 °C were more leaky than the ones annealed at 600 °C. This could be correlated with the dielectric constants of the samples annealed at 550 °C and 600 °C. The values for 550 °C annealed samples were slightly higher than the ones for 600 °C. In addition, it can be seen that I-V patterns of crystalline $SrTiO_3$ layers, annealed in different atmospheres have different shapes. The films, annealed in oxygen atmosphere possessed lower leakage currents (Fig. 3a) at lower voltages; however the values are comparable at higher voltages. This different behavior indicates that the annealing conditions influence the structure of the $SrTiO_3$ and therefore impacts the leakage mechanisms at different electric fields.

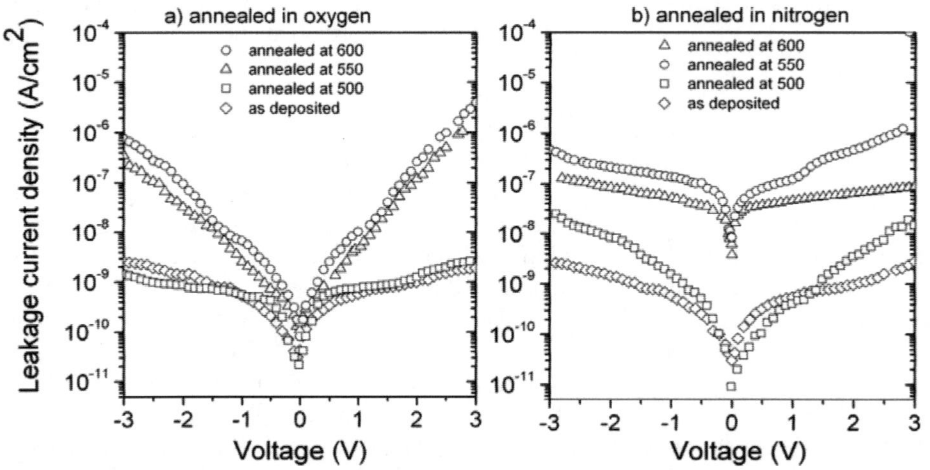

Fig. 3. Leakage current densities as a function of voltage of $SrTiO_3$ based MIM capacitors for the as deposited samples (◊), annealed at 500 °C (□), 550 °C (○) and 600 °C (Δ) samples in a) oxygen and b) in nitrogen atmospheres.

Finally, leakage current density of the Pt/$SrTiO_3$ system is compared with the TiN/$SrTiO_3$ case in Fig. 4. A clear difference of around 4 orders of magnitude is observed if Pt is used instead of TiN, although the thicknesses and the deposition conditions for $SrTiO_3$ were identical in this case.

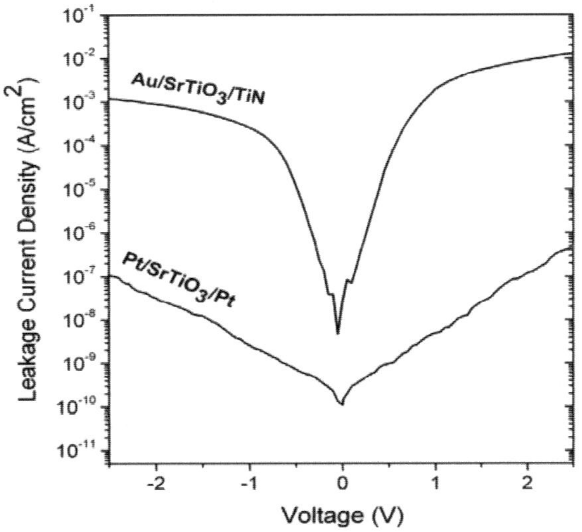

Fig. 4. Leakage current densities as a function of voltage of crystalline $SrTiO_3$ deposited on Pt and on TiN bottom electrodes.

Another important parameter that can be extracted from the C-V measurements is the quality factor, since no dielectric is perfect and some losses, called *tanδ* or dissipation factor of the capacitor must be recognized. High performance MIM devices must have low losses or high *1/tanδ* ratio which is called the quality factor. The comparison of quality factors for $SrTiO_3$ annealed at different temperatures and atmospheres is presented in Fig. 5. For the as deposited samples the quality factors are very high and values ~ 600 have been reached. However, the quality factors decreased to ~ 20 if the annealing temperature of 550 °C was applied independent of the annealing atmosphere. This is related to the fact that at lower frequencies the quality factor is mainly influenced by the resistance of the dielectric itself. This observation is in the agreement with the I-V measurements, which are plotted in Fig. 3a and Fig 3b, where lower leakage currents for crystalline samples were observed in comparison with as deposited samples. The lower leakage currents correspond to the higher resistance of the $SrTiO_3$ layers and therefore influence the quality factor of $SrTiO_3$ based MIM capacitors.

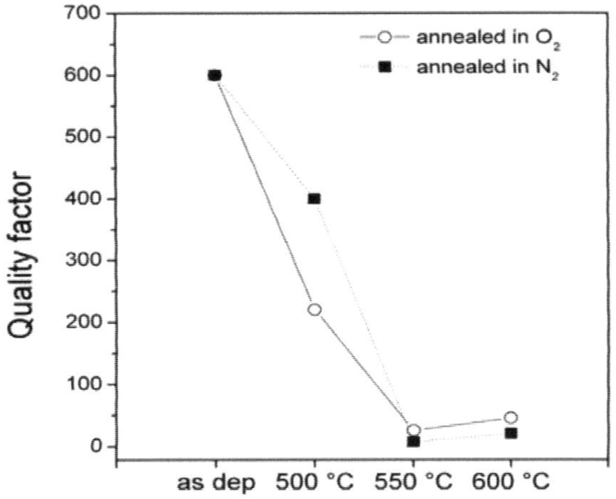

Fig. 5. Quality factor dependence on the annealing temperature for the $SrTiO_3$ samples annealed in oxygen (o) and in nitrogen (■) atmosphere.

A key characteristic regarding the performance of MIM capacitors is the capacitance voltage linearity, which indicates the dependence of capacitance on the applied voltage. Especially for analog applications, the variation of capacitance in the device must be as low as possible with respect to applied voltage. This stability of capacitance variation is commonly described by the following polynomial law [11]:

$$C(V) = C_0(\alpha V^2 + \beta V + 1) \quad (1)$$

where C_0 is the capacitance at zero bias voltage, and α and β represent the quadratic and the linear capacitance coefficients, respectively. Experimentally, this variation, or the alpha parameter, can be extracted from standard C-V curves after fitting the experimental data with the Eq. 1. The effect is visualized in Fig. 6 and Fig 7, where the normalized capacitances (C/C_0) of $SrTiO_3$ based MIM capacitors are plotted as a function of the applied voltage. In general, the extracted quadratic nonlinearity coefficients for as deposited and at 500 °C annealed (both in oxygen and nitrogen) samples were rather small (30 and 41 ppm/V^2) and fulfilled the requirement of <100 ppm/V^2 defined by ITRS. However, if $SrTiO_3$ was annealed at 550 °C or 600 °C, which is the temperature of crystallization of $SrTiO_3$, the coefficient alpha changed from the positive values to the negative values and reached -2700 ppm/V^2 after 550 °C anneal, and then stabilized to -1790 ppm/V^2 after 600 °C anneal in nitrogen atmosphere. In oxygen atmosphere, the corresponding values were -645 ppm/V^2 after 550 °C anneal and -387 ppm/V^2 after 600 °C anneal. This difference is also probably influenced by the structural changes of $SrTiO_3$ after annealing it at different conditions. Indeed, the oxygen content in the films as well as oxygen vacancies have been reported to be the origin of the nonlinearities observed in MIM capacitors [12, 13]. On the other hand, the mechanism for nonlinearity is rather complicated. Literature reports propose that the alpha coefficient could be related to the dielectric constant and elastic properties of the insulator [14], interface reactions between the electrode and dielectric [15] or carbon content in the films [16] to name a few. In addition, this transition from positive to negative α values of $SrTiO_3$ was also observed in the literature [12], where the authors observed that as deposited amorphous samples possessed also positive α values of 2200 ppm/V^2, whereas the annealed crystalline samples had strong negative α values (-12800 ppm/V^2). The authors claimed that crystalline $SrTiO_3$ is paraelectric material and is known to display negative voltage coefficients. In this study, the values were much lower, but the tendency was also similar.

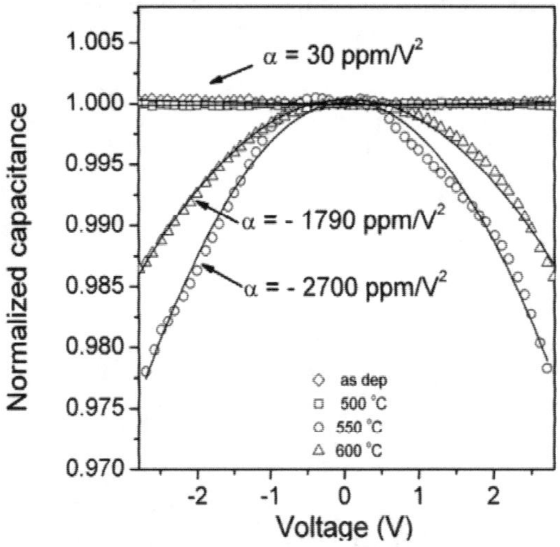

Fig. 6. Normalized capacitance density as a function of applied voltage for as deposited (◊) and in nitrogen annealed $SrTiO_3$ samples: 500 °C (□), 550 °C (○) and 600 °C (Δ). The black solid lines represent eq. (1).

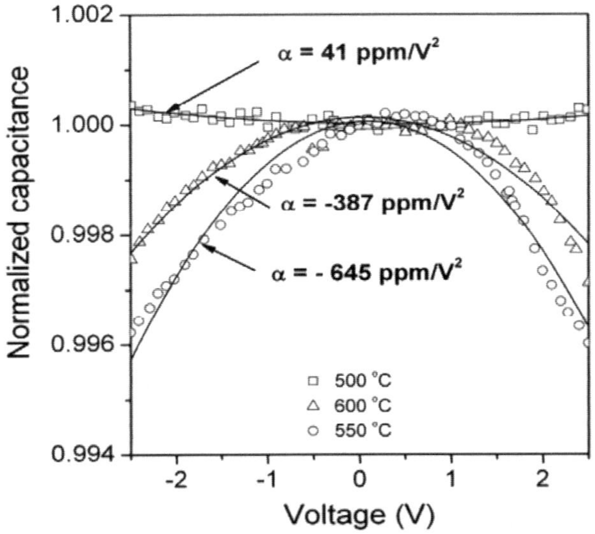

Fig. 7. Normalized capacitance density as a function of applied voltage for $SrTiO_3$ layers, annealed in oxygen atmosphere at 500 °C (□), 550 °C (○) and 600 °C (Δ). The black solid lines represent eq. (1).

4. Conclusions

Metal-Insulator-Metal capacitors with ALD deposited $SrTiO_3$ films have been investigated in this study. Crystallization and electrical properties were investigated for the as deposited as well as in oxygen or nitrogen at 500-600 °C annealed samples. The crystallization temperature of 550 °C for $SrTiO_3$ was detected from the XRD measurements independent from the annealing atmosphere. The crystallized $SrTiO_3$ films had high dielectric constant of ~ 85 and relatively low leakage current densities (~ 10^{-6}-10^{-7} A/cm^2) at ± 2V. On the other hand, the extracted quadratic capacitance-voltage nonlinearity coefficients for crystalline $SrTiO_3$ were higher than the ones defined by ITRS, and were different for the samples annealed in nitrogen or oxygen at 550 °C (-1790 ppm/V^2 and -387 ppm/V^2). Nevertheless, $SrTiO_3$ remains one of the most promising alternative dielectric for MIM applications.

References

[1] Radio Frequency and Analog/Mixed-Signal Technologies for Wireless Communications, International Roadmap for Semiconductors (Semiconductor Industry Association, Palo Alto 2011 update)
[2] Yu X et al 2003 *IEEE electron Device Lett.* **24** 63
[3] Tu Y et al 2003 *VLSI Tech. Dig.* 79
[4] Goux L et al 2006 *J. Electrochem. Soc.* **153** F132
[5] Chiang K et al 2005 *Electron Device Lett.* **26** 728
[6] Chiang K et al 2007 *J. Electrochem. Soc.* **154** H214
[7] Kim S et al 2010 *Adv. Func. Mater.* **30** 715
[8] Piskunov S et al 2004 *Comput. Mater. Sci.* **29** 165
[9] Lukosius M et al 2012 *J Solid State Sci. and Technol.* **1** N1
[10] Van Elshocht et al 2009 *J Vac. Sci. Technol. B* **27** 209
[11] Ding S et al 2003 *Electron Dev Lett.* **24** 730
[12] Jorel C et al 2009 *Appl. Phys. Lett.* **94** 253502
[13] Gonon P et al 2007 *Appl. Phys. Lett.* **91** 172909
[14] Wenger Ch et al 2008 *J. Appl. Phys.* **103** 104103
[15] Wenger Ch et al 2009 *Thin Solid Films* **517** 6336
[16] Miao B et al 2008 *J. Appl. Phys* **104** 054510

Fuzzy-logic-based approach to study the electrons mobility in nanoscale Double Gate MOSFETs

T. Bendib[1], F. Djeffal[1,2], D. Arar[1], Z. Dibi[1] and A. Ferdi[1]

[1] LEA, Department of Electronics, University of Batna, Batna 05000, Algeria

[2] LEPCM, University of Batna, Batna 05000, Algeria

E-mail: bendib05.t@gmail.com, faycaldzdz@hotmail.com

Abstract. The Double Gate silicon (DG) MOSFET with extremely short-channel length has the appropriate features to constitute the devices for nanoscale circuit design. To develop a physical model for extremely scaled DG MOSFETs, the electron mobility distribution throughout the channel must be accurately determined under the application of drain and gate voltages. However, the modeling of the transport behavior for the nanoscale structures requires the use of overkill methods and models in terms of complexity and computation time (self-consistent, quantum computations,..). Therefore, new methods and techniques are required to overcome these constraints. In this work, a new approach based on the fuzzy logic computation is proposed to model the electron mobility behavior for nanoscale DG MOSFETs. The proposed approach can be implemented into devices simulator to study the nanoscale CMOS-based circuits.

1. Introduction

These In recent years, Double Gate (DG) MOSFET has attracted a lot of attention for nanoscale CMOS circuit design. This is mainly due to the superior control of short channel effects (SCEs). The advantages advocated for DG MOSFET include: light doping of the channel reducing the mobility degradation due to the elimination of impurity scattering; good control of short channel effects; ideal subthreshold swing due to the elimination of substrate doping; etc [1–3]. Although the operation of DG MOSFET is similar to the conventional MOSFET, the transport mechanism in the nanoscale DG MOSFET is more complicated. In addition, physical phenomena, such as the quantum mechanical and short channel effects, have to be considered. Therefore, simulation tools which can be explored to design nanoscale CMOS circuits require new approaches and modeling techniques that capture the physics of nanoscale quantum transport accurately to guide the design for nanoscale CMOS-based devices. Since electrons mobility is one of the key parameters defining nanoscale DG MOSFET performance, reliable modeling techniques for its evaluation in nanoscale devices are of prime importance. To extract information accurately about the electrons mobility require the solution of Schrödinger and Poisson equations (SPE), assuming quantum effects are to be fully accounted [4]. But from the nanoscale CMOS-based circuit design point of view even 2-D solution of numerical SPE is an overkill approach in term of both complexity and computational cost [4]. For analytical device modeling, the Gradual Channel Approximation (GCA) model is widely used for analytical

[1] To whom any correspondence should be addressed.

IOP Conf. Series: Materials Science and Engineering **41** (2012) 012016 doi:10.1088/1757-899X/41/1/012016

semiconductor device modeling. However, it is well known that GCA model has not accuracy enough for simulation of nanoscale devices in which, rapid changes and confinement of electric field, and non-local effects such as velocity overshoot and quantum effects are pronounced [5-7]. Accurate analytical mobility models are required to be utilized in nanoscale circuit simulators.

In the present paper, we present an efficient technique for nanoscale DG MOSFETs modeling, where simple and accurate mobility model can be automatically achieved from a computational process to maximally reduce human trial and error efforts. The developed model should be able to correctly represent the electrons mobility for the nanoscale device, and effectively capture the electrical behavior of the device due to process variations. In this context, in order to achieve the required accuracy and model simplicity, in this paper we present the applicability of Fuzzy Logic (FL) computation approach to develop an analytical electron mobility model for nanoscale DG MOSFETs. Moreover, the model should be formulated such that it can be conveniently incorporated into existing circuit simulators for high-level circuit simulation and yield design.

2. Modeling methodology

The basic structure of the DG MOSFET investigated in this study is shown in Fig.1. As MOSFET feature size moves into nanoscale regime, canonical carrier transport theories are no longer capable of describing carrier transport accurately. The canonical theories are basically derived from the Boltzmann transport equation (BTE), with more or fewer approximations being made [8]. These models focus on scattering-dominant transport, which typically occurs in long channel devices. However, transistors operate in ballistic or quasi-ballistic transport regime. Simulations using conventional models may either under predict or over predict the electrical device behavior [8-10]. Fig. 1 shows the cross-sectional structure of the DG MOSFET investigated in this work.

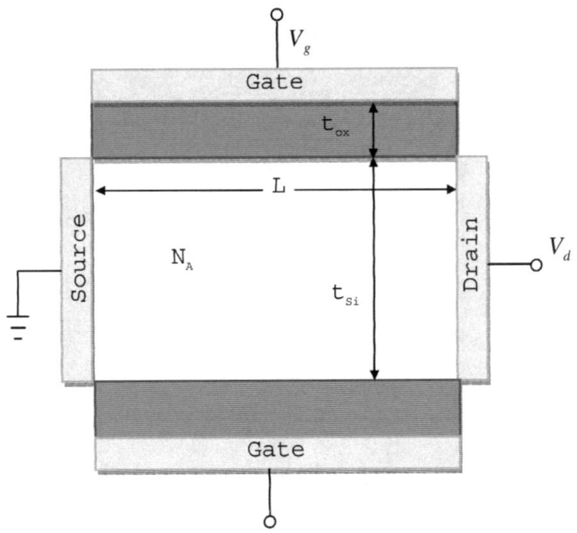

Figure 1. Schematic sketch of symmetrical DG MOSFET structure investigated in this study with (channel doping N_A=10^{15} cm^{-3}, t_{si} represents silicon thickness and t_{ox} is the oxide thickness).

The substrate is p-type Silicon with a thin channel thickness, and doped concentration is 10^{15} cm^{-3}. The source and drain are of n-type with doped concentration of 10^{20} cm^{-3}.

The design of optimal DG MOSFET will require new insights into the underlying physics, especially electrons transport in the channel region. Quantum-mechanical confinement of electrons significantly affects the electrical characteristics of highly scaled MOSFETs. Therefore, an analytical DG MOSFETs modeling with arbitrary Si-film thickness (less than 5 nm) is needed for physical insight as well as for a reliable device model. To simulate nanoscale DG MOSFETs, the Non-Equilibrium Green's Function Formalism (NEGF) provides one of the best frameworks available [4,11]. Based on the efficiency proven by NEGF for the modeling of nanoscale DG MOSFETs, we have used the NEGF formalism to form the database which will be used to learn our FL-based model as it is shown in Fig.2.

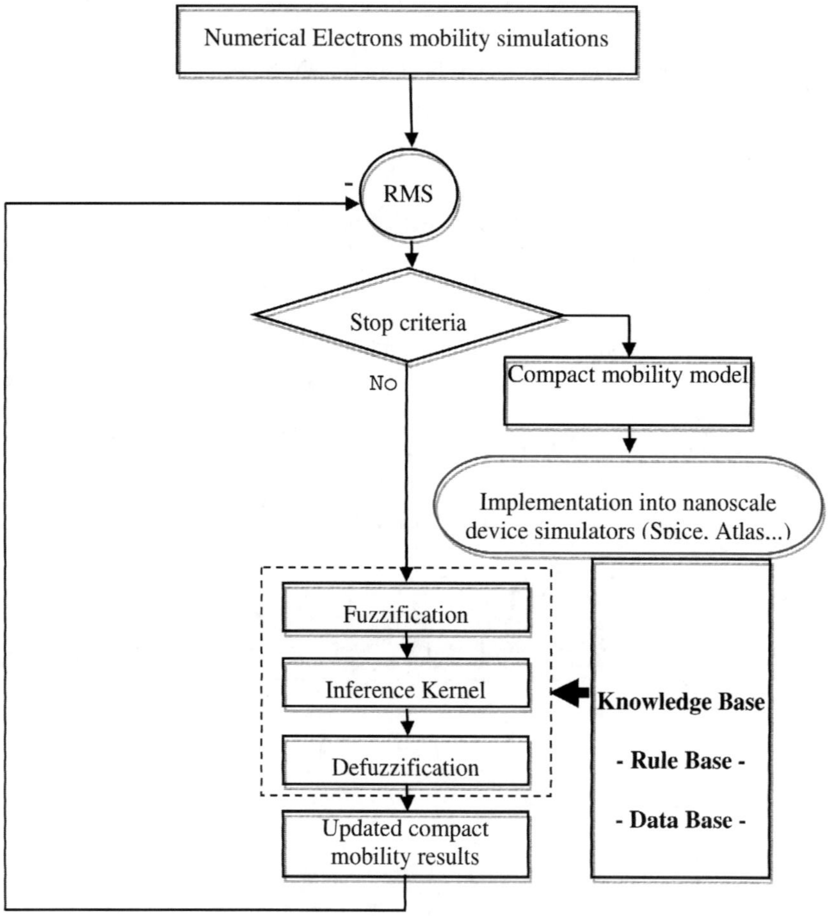

Figure 2. Flowchart of our FL-based electrons mobility computation approach.

The FL-based modeling can provide an effective representation of complex non-linear systems in terms of fuzzy sets and fuzzy inference [12]. Several published works [12-14] offer extensive explanations to the fundamentals of FL and its wide range of applications. The input–output behavior of a fuzzy system is programmable using linguistic information in the form of *IF* (preconditions) *THEN* (postconditions) rules, describing an approximate or qualitative knowledge of an observed process. FL is a superset of conventional Boolean logic that has been extended to include the concept of partial truth-truth-values between 'completely true' and 'completely false'. Due to its simple

mechanism and high performance for behavior modeling, FL can be applied to study the electrons mobility behavior for nanoscale transistors, which is the main objective of this work.
The first step of our approach consists of the compact analytical electrons mobility model for conventional long channel DG MOSFETs. For the long channel transistor with thick silicon body, the low-field mobility, μ_0, does not change with thickness and therefore, the study of the electron transport becomes in this case simple. However, the behavior of the effective electron mobility for nanoscale transistors tends to decrease [15]. In this case, the in-depth average values of the electron mobility and the effective electric field weighted by the electron distribution are given [16] by

$$\mu_{eff} \approx \frac{\mu_0}{1+\left(\frac{E_{eff}}{E_0}\right)^{\delta}} = \frac{\mu_0}{1+\eta\frac{Q_{iq}}{2\varepsilon_{si}E_0}} \qquad \text{with } E_{eff} = \frac{Q_{iq}}{2\varepsilon_{si}} \tag{1}$$

where μ_0 is the silicon low-field mobility ($\mu_0 = 500cm^2/V.s$) E_0 is a fitting parameter which can be presented by an applied electric field, which mainly depends on V_{gs}. The value of this latter will be calculated using the FL-based computation ε_{si} is the silicon permittivity, η is a fitting parameter ($\eta = 10^{-3}$) and E_{eff} represents the applied effective electric field in vertical direction, which mainly depends on the inversion charge distribution in the channel. Based on the inversion charge model proposed by *Taur et al* [16], a new extended model, Q_{iq}, called the corrected inversion charge model of undoped DG MOSFET, in which the short channel effects have been introduced as fitting parameters. This latter is given as

$$Q_{iq} = C_{oxq}\left[-2C_{oxq}\frac{V_{th}^2}{Q_0} + \sqrt{\left(2C_{oxq}\frac{V_{th}^2}{Q_0}\right)^2 + 4V_{th}^2\ln\left(1+e^{\frac{V_{gs}-V_{fb}-\phi_F+V_T+\delta V_{Tq}}{2V_{th}}}\right)}\right] \tag{2}$$

where V_{fb} is the flat band voltage, V_T is the threshold voltage, and Q_0 is the charge coefficient given [16] by $Q_0 = (8V_{th}\varepsilon_{si}/t_{si})$. C_{oxq} is the quantum oxide capacitance, which is given as function of channel thickness, $C_{oxq} = \frac{C_{ox}}{1+\xi_1 C_{ox} t_{si}}$. V_{Tq} represents the quantum threshold voltage, which includes the channel thickness effect. This variation can be given as, $V_{Tq} = V_T + \delta V_{Tq}$.
with δV_{Tq} represents the threshold voltage shift due to the quantum effects, and it is given by, $\delta V_{Tq} = \frac{\xi_2}{t_{si}^2}$. ξ_1 and ξ_2 are fitting parameters which will be determined using our FL-based approach as it shown in Fig.2.
From Eq. 1, an accurate modeling of the inversion charge concentration leads to an accurate electrons mobility computation. So, in our FL-based computation we focus to study the inversion charge behavior.
To define the fuzzy associative memory, we need some knowledge about how each of these parameters affects the mobility behavior:

- V_{Tq} shifts the whole $Q_{iq} - V_{GS}$ curve.
- ξ_1 describes the curvature of $Q_{iq} - V_{GS}$ and provides a scale factor.
- ξ_2 provides a scale factor.

Based on the effect of each parameter on the $Q_{iq} - V_{GS}$ characteristics, we can define the following rules in order to develop our knowledge Base (Fig.2):

- If the calculated curve $Q_{iq} - V_{GS}$ is to the right of the numerical data, then decrement V_{Tq}, and vice versa.

IOP Conf. Series: Materials Science and Engineering **41** (2012) 012016 doi:10.1088/1757-899X/41/1/012016

- If the calculated curve $Q_{iq} - V_{GS}$ is more curved than the numerical data, then decrease ξ_1, and vice versa.
- If the calculated curve $Q_{iq} - V_{GS}$ is over the numerical data, then decrease ξ_2, and vice versa.

In order to implement the above rules, we have used the triangular fuzzy sets, while defuzzification is carried out through the method of centre of area. The input and output parameters are normalized by using the numerical database as reference. The linguistic variables chosen for this FL-based model are the principal and secondary errors (Er_P and Er_S) for each parameter of the compact electron mobility model. These errors can be calculated, for each parameter, from:

$$Er\_P(\xi_1) = \frac{Q_{iq-N}(\mathrm{V_{GS}} \approx \mathrm{V_{Tq}})}{Max(Q_{iq-N})} - \frac{Q_{iq-F}(\mathrm{V_{GS}} \approx \mathrm{V_{Tq}})}{Max(Q_{iq-N})} \tag{3a}$$

$$Er\_P(\xi_2) = \frac{Q_{iq-N}(\mathrm{V_{GS}} \approx \mathrm{V_{sat}})}{Max(Q_{iq-N})} - \frac{Q_{iq-F}(\mathrm{V_{GS}} \approx \mathrm{V_{sat}})}{Max(Q_{iq-N})} \tag{3b}$$

$$Er\_S(\xi_1,\xi_2) = \frac{Q_{iq-N}(\mathrm{V_{GS}} \approx \mathrm{V_{M}})}{Max(Q_{iq-N})} - \frac{Q_{iq-F}(\mathrm{V_{GS}} \approx \mathrm{V_{M}})}{Max(Q_{iq-N})} \tag{3b}$$

where Q_{iq-N} and Q_{iq-F} represent the 2D numerical inversion charge simulation and FL-based inversion charge model, respectively. V_M is the medium voltage given by $V_M = \frac{V_{Tq} + V_{DD}}{2}$ with V_{DD} represents the supply voltage.

The both errors are the input linguistic variables and the inversion charge is the final output linguistic variable. In this work, the principal error of each parameter represents the drain current deviation affected, only, by the main parameter, and the secondary one is the drain current deviation affected by a combination of several parameters (mutual effect). In this approach, we design a fuzzy inference system based on the past known behavior of a target system. The fuzzy system is then expected to be able to reproduce the behavior of the target system (Fig.2). Each of the input and output fuzzy variables is assigned seven linguistic fuzzy subsets varying from negative large (NL) to positive big (PL). Each subset is associated with a triangular membership function to form a set of seven membership functions for each fuzzy variable. The linguistic terms chosen for this controller are seven. They are negative large (NL), negative medium (NM), negative small (NS), zero (Z), positive small (PS), positive medium (PM) and positive large (PL). After assigning the input, output ranges to define fuzzy sets, mapping each of the possible five input fuzzy values of principal error(Er_P), three input fuzzy values of secondary error (Er_S) to the seven output fuzzy values is carried out through a rule base. Table.1 shows the fuzzy associate memory table of the fuzzy $V_{Tq}(\xi_1)$ and ξ_2 controllers.

Table 1. Fuzzy associate memory table (FAM) for the fuzzy $V_{Tq}(\xi_1)$ and ξ_2 controllers.

		Principal error Er_P				
		NL	NM	Z	PM	PL
Secondary error Er_P	N	NL	NM	NS	PS	PM
	Z	NM	NM	Z	PM	PM
	P	NM	NS	PS	PM	PL

3. Results and discussion

From Fig.2, the computation will iterate from an initial point until a stopping condition is found. For each iteration, new values for the parameters will be estimated by the fuzzy calculator. The global RMS (Root Mean Square) error between the numerical and the calculated results, considering all points on the database, will be updated. In our case, we have used a stopping condition of global RMS error less than a 5%. Fig. 3 shows good agreement between numerical and predicted results of the electron mobility for the investigated nanoscale DG MOSFETs. It is to note that our simulations were carried out for a wide range of nanoscale DG MOSFET parameters (geometrical and electrical parameters) in order to ensure the applicability of the proposed FL-based model to study the electron mobility for the nanoscale DG MOSFETs.

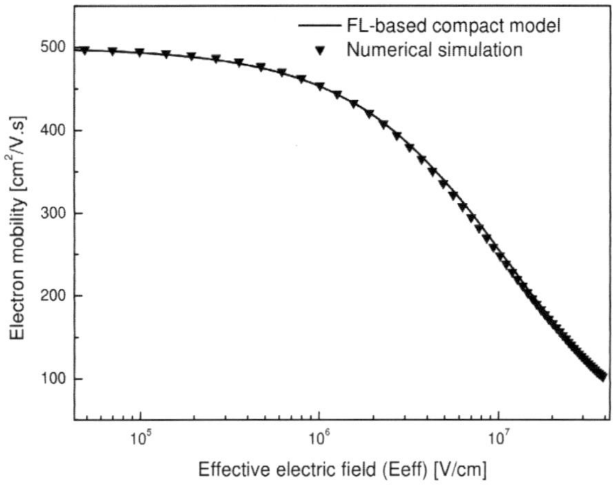

Figure 3. Electrons mobility characteristic calculated from the FL-based compact model (solid lines), compared with numerical simulations (symbols) for $t_{si} = 5\ nm$ *and* $t_{ox} = 1.5\ nm$.

Table 2 gives a comparison of the CPU (central processing unit) time requirements for the simulation of the electrons mobility with various approaches where the FL-based model computation time should be compared to the orders of magnitude increase in computation time for more rigorous electrons mobility models, such as those based on the manual parameters computation. Obtained results can be explained by the fact that the FL-based techniques are characterized as computational models based on parallel distributed processing of data. Therefore, the fuzzy computation provides practical insight into electrons transport modeling in nanoscale devices without the uncertain accuracy or meticulous tuning effort that face more rigorous mobility models. The FL-based modeling is a step towards a new generation of simulation tools that will allow device and material engineers to explore new classes of electronic devices.

Table 2. Comparison between the various approaches of modeling of the electrons mobility in nanoscale DG MOSFET.

	Computation time	Accuracy/ approach complexity
Numerical simulations [4]	Hours	Accurate / complex
Analytical computation [7]	Seconds	Not accurate / simple
Our FL-based computation	Seconds	Accurate / simple

Figure.4 shows the evolution of the Fuzzy logic parameters versus the simulation cycle times using the computer (P.IV-3GHz). It is clearly shown that the convergence point of the Fuzzy computation process can be achieved for 1×10^4 cycle times (several seconds), for which the stability of the generated FL parameters was found. This result makes the fuzzy logic-based approach as a powerful tool to study the nanoscale devices.

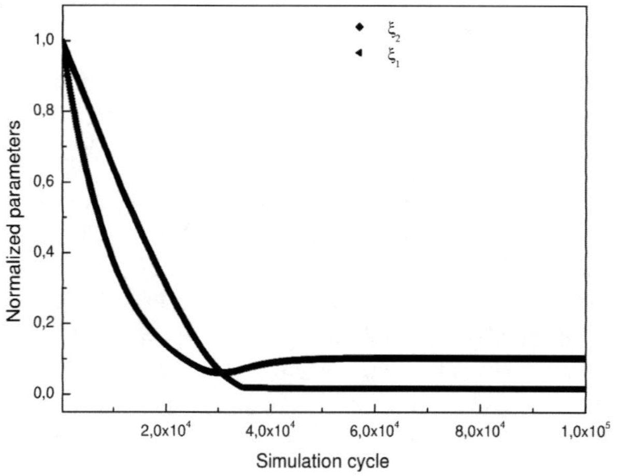

Figure 4. Evolution of the Fuzzy logic parameters versus the simulation cycle times.

4. Conclusion

In this paper, we showed the applicability of the FL technique to study the electron mobility in nanoscale DG MOSFETs. The parameters extraction of the extended electron mobility model was completed in a relatively short time, with no need for user intervention during the search. The proposed approach can be extended to include other complex phenomena in nanoscale devices like: aging phenomena, hot-carriers, and length modulation effect, which cannot be modelled using conventional analytical investigations. The encouraging comparisons between numerical simulations and our compact model results have indicated that the proposed approach is particularly suitable to be incorporated in electronic circuit simulators to study the nanoscale CMOS devices.

IOP Conf. Series: Materials Science and Engineering **41** (2012) 012016 doi:10.1088/1757-899X/41/1/012016

5. References

[1] International Technology Roadmap for Semiconductors (ITRS), Published online at http://public.itrs.net, 2009.
[2] Djeffal F, Bendib T and Abdi M.A 2011 *Microelectronics. J.* **42** 1391
[3] Barin N, Braccioli M, Fiegna C and Sangiorgi E 2007 *IEEE Trans. Nanotechnology* **6** 421
[4] Venugopal R, Ren Z, Datta S and Laundstrom M.S 2002 *J. Appl Phys* **92** 253
[5] Stern F 1972 *Phys. Rev. B* **5** 4891
[6] Baccarani G and Reggiani S 1999 *IEEE Trans. Electron Device* **46** 1656
[7] Chen Q, and Meindl J.D 2004 *Nanotechnology* **15** 8549
[8] Reyboz M, Rozeau O, Poiroux T, Martin P and Jomaah J 2006 *Solid-State Electronics* **50** 1276
[9] Iñiguez B, Fjeldly T. A, Lázaro A, Danneville F, and M. Jamal Deen 2006 *IEEE Trans. Electron Device* **53** 2128
[10] Xiong S, King T. J and Bokor J 2005 *IEEE Trans. Electron Device* **52** 1859
[11] Djeffal F, Chahdi M, Benhaya A and Hafane M.L 2007 *Solid-State Electron* **51** 48
[12] Yager RR and Zadeh LA, *Norwell MA: Kluwer* 1991.
[13] Mishra K, Sarma I. G, and Swamy K. N 1994 *J. Guid. Control.Dyna* **17** 1389
[14] Esposito M and Pietro G. D 2011 *Engineering Applications of Artificial Intelligence* **24** 1340
[15] Reggiani S, Gnani E, Gnudi A, Rudan M and Baccarani G 2007 *IEEE Transactions on Electron Devices* **54** 2204
[16] Taur Y 2001 *IEEE Trans. Electron Devices* **48** 2861

IOP Conf. Series: Materials Science and Engineering **41** (2012) 012017 doi:10.1088/1757-899X/41/1/012017

Multilevel current stress technique for investigation thin oxide layers of MOS structures

V V Andreev[1,4], G G Bondarenko[2], V M Maslovsky[3], A A Stolyarov[1]

[1] Bauman Moscow State Technical University, Kaluga Branch. 2, Bazhenov St., Kaluga, 248600, Russia.
[2] Moscow state institute of electronics and mathematics (Technical University), 12 Malaya Pionerskaya, Moscow, 113054, Russia.
[3] The state unitary enterprise of a city of Moscow Research-and-production centre "SPURT", Zelenograd, West of the 1-st proezd 4, 124460, Russia.

E-mail: andreev@bmstu-kaluga.ru

Abstract. In this study, a new technique of multilevel current stress for investigation thin oxide layers of MOS structures is proposed. This technique allows to investigate the generation and relaxation of positive and negative charges, accumulating in, nano-thickness gate dielectric of MOS structures under many stressing situations. The parameters characterizing the change of charge state in the thin oxide layers of MOS structures during the stress have been monitored by means of time dependence of voltage shift applied to a sample during the injection. The method takes into account the process of the MOS capacitance charging by the constant current pulse. In comparison with the conventional techniques our method is nondestructive and enables to reduce the high-field stress time while measuring the injection current-voltage characteristics of the MOS-structure. We consider this method to provide higher accuracy and to decrease the probability of dielectric breakdown. The application of present technique was carr mied out during the investigation of charge generation and relaxation, during and after high-field tunnel injection of electrons in thin thermal film of SiO_2. The thin oxide layers of MOS structures after plasma and irradiation treatments also are investigated.

1. Introduction

Generation and following relaxation of positive and negative charge in thin gate dielectric of MOS structures, observed under different stressing situations (such as high-field charge injection, exposure to ionizing radiation, plasma treatment etc.) are the most important degradation mechanism of VLSI devices and is investigated intensive at present time [1-5]. Though there is a lot of literature dedicated to this matter, there is no complete understanding of expression about physical phenomena taking place at that. So, as a possible mechanism of positive charge generation, following processes are studied: impact ionization in SiO_2 film produced by injected electrons at high-fields and leading to holes generation [4], hole injection from anode [5], redistribution of hydrogen [1,3] and so on. To explain after-stress relaxation of positive charge, different models are used also, for example, hydrogen transportation model [4], trapped hole conversion [3], and others. Such a variety of used

[4] E-mail: andreev@bmstu-kaluga.ru

models is specified by not complexity of phenomena taking place only, but by lack of experimental data. It is impossible to obtain these data without new investigation techniques allowing to obtain new information about charge generation and relaxation in MOS structures.

In this study, a new technique of multilevel current stress for investigation thin oxide layers of MOS structures is proposed. This technique allows to investigate the generation and relaxation of positive and negative charges, accumulating in, nano-thickness gate dielectric of MOS structures under many stressing situations.

2. Technique description

Usually using the technique of the charge injection into the MOS-structure dielectric with direct current, the experimental data analysis is started after all the current flowing through the MOS-structure becomes the injection current. The processes in the MOS-structure during MOS-structure capacitance charging are as a rule neglected. This considerably reduces the usefulness of such a technique, as will be shown below.

The proposed injection technique consists in applying a constant current pulse to the MOS-structure and measuring the time dependent voltage, $V_I(t)$. The data analysis is carried out taking into account the MOS-structure capacitance charging. To explain the principle of the technique we will consider the diagrams for different currents and voltage on the MOS-structure during the measurements (Fig. 1). The variation of $V_I(t)$ during the MOS-structure capacitance charging can be conventionally divided into the following ranges: the low-voltage and the high-voltage.

The low-voltage range is considered to be a range, over which no charge is injected into the dielectric and all the current flowing through the sample only is the capacitive. In the high-voltage range, in addition to the capacitive current, exists the current caused by charge injecting into the dielectric. The recording of the MOS-structure capacitance charging process over the low-voltage range using the constant current technique allows to test the MOS-structure parameters, which are determined, as a rule, with the C-V technique. In this case it is necessary to charge the MOS-structure by a constant current pulse with an amplitude I_1 (fig. 1 a) from the mode of an accumulation or deep inversion so that the Fermi level passes over all the band gap part which is interesting for us (fig. 1, portion 2). As the condenser capacity is charged by a direct current taking into account the equation of a neutrality of a charge time dependence of capacity of MOS-structure on this portion appears inversely proportional a derivative $dV_I(t)/dt$ [6-10].

Using C(t) and taking into account $V_I(t)$, it is possible to determine the C-V curve which is the low-frequency C-V dependence (Fig. 1, portion A). The requirements for the constant current pulse amplitude in the proposed technique are similar to the requirements for the current amplitude in the quasistatic C-V technique [3,6,9]. With a further charging structure in this portion flowing through the dielectric current, will be capacitive, and the capacity of the MOS structure is constant and equal to the dielectric capacitance C_i).

Taking into account that $V_I(t)$ is functionally related to the surface potential, the charge for the MOS structure constant current (fig. 1 b, portion 2) the time dependence of the surface potential can be determined as follows:

$$\varphi_s(t) = V_I(t) - \frac{I_2}{C_i} \cdot t + A, \tag{1}$$

where A is the constant of integration. The integration constant A, as well as in the low-frequency C-V technique, is defined by the initial conditions or from other measurements.

Taking into account both $\varphi_s(t)$ and $V_I(t)$, we obtain the dependence $\varphi_s(V_I)$, which allows to determine the flat band voltage. As a result, using the above technique can be determined depending on: the low-frequency capacitance of the MOS structure bias voltage, the surface potential of the voltage and the density of surface states on the surface potential is traditionally produced using low-frequency C-V technique.

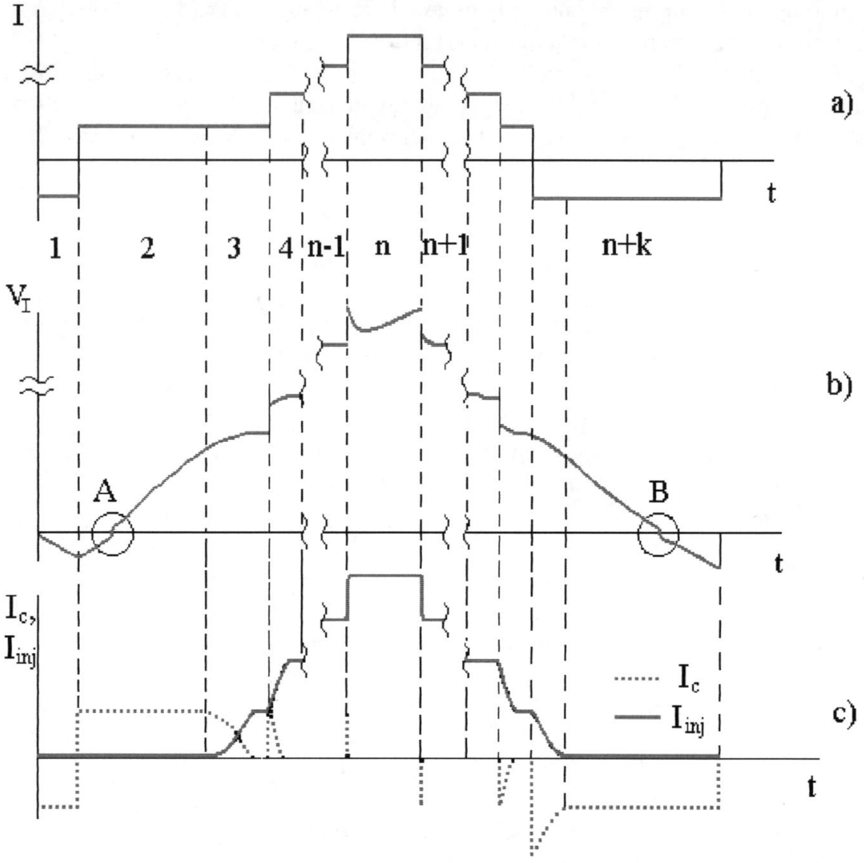

Figure 1. The time dependences of the values corresponding to: the amplitude of the current pulse applied to the MOS-structure (a), the measuring voltage on the MOS-structure (b), the capacitance and injection current flowing through the MOS-structure (c). A and B are portions at the V_I depending where are determined the C-V characteristics.

In portion 3 (Fig. 1, b) all the current flowing through the dielectric remains to be the capacitive one; the MOS-structure capacitance becomes to be constant and equal to the dielectric capacitance C_i. The dielectric capacitance measurement in this portion allows to determine the dielectric thickness when its dielectric constant is known.

Over the high-voltage range (Fig. 1, b, portion 3), when starting the charge injection into the dielectric (Fig. 1, c), the recording of the MOS-structure capacitance charging process also allows to expand the possibilities of the constant current technique [5]. Using the constant current technique in the high-field injection range the condition in which the dielectric charge variation during the high-field injection is much less than the value of the injected charge Q_{inj} should be satisfied. When applying a rectangular current pulse to the studied sample it is possible to express the MOS-structure charge neutrality equation as the following:

$$Q_3 = Q_c + Q_{inj}, \qquad (2)$$

where Q_3 is the charge supplied to the sample ($Q_3 = I_3 \cdot t$), Q_c is the capacitive component of the MOS-structure charge.

Solving Eq. (2) with respect to Q_{inj} and considering that an MOS-structure capacitance under the high-field injection is equal to the dielectric capacitance, the expression for determining the value of the charge injected into the dielectric has the following form:

$$Q_{inj}(t) = \int I_{inj}(t)dt = I_3 \cdot t - C_i \cdot V_I(t), \tag{3}$$

where I_{inj} is the injection current. Differentiating Eq. (3) with respect to time, the expression for determining the injection current has the following form:

$$I_{inj}(t) = I_3 - C_i \cdot \frac{dV_I(t)}{dt}. \tag{4}$$

Measuring the time-dependence $V_I(t)$ and using Eq. (4), we determine the experimental I-V dependence for the studied sample under the high-field stress. The minimum current in the I-V dependence is limited by the measurement accuracy, and the maximum level is limited by the value of I_3.

The given way of measurement I-V of the characteristic is expedient for using in that case when it is necessary to minimize amount of the injected charge. If carry out research of change of charging characteristics of MOS-structures wider range of high-fields carry out step increase in amplitude applied current (fig. 1, portions 4..., n-1, n; where n - portion number on which the basic injection of a charge is made), and current-voltage characteristics receive on values of a current and voltage on each step. The quantity of steps is defined by demanded accuracy of reception I-V and finite value of a current at which charge injection in dielectric will be spent. Duration of a step should provide a full charge of capacity of MOS-structure and transition in a mode of injection of a charge. The equation (4) can be used also in need of expansion of a range measured I-V towards currents less I_3.

For increase of accuracy of definition of value of the injected charge it is expedient to use the following equation:

$$Q_{inj}(t) = \sum_{j=1}^{n} \left[I_j \cdot \Delta t_j - C_i \cdot V_{Ij}(t) \right], \tag{5}$$

where *j* - portion number, *n* - number of portions, Δt_j – time duration of current pulse on *j*-th a portion. Application of the equation (5) gives the chance to consider processes charging and discharging of capacity of MOS-structure and, thereby, to raise accuracy of definition of value of the injected charge.

On n-th portion (fig. 1) is carried out injection in dielectric the demanded size of the charge, allowing to study processes of charging degradation of MOS-structures in high-fields. Then portion reduction of amplitude applied current is spent (fig. 1, portions n+1..., n+k). As a result it is possible to measure *I-V* the characteristic right after high-field stress and to receive full picture of charging degradation dielectric, including influence quickly relaxing charges which flow down after removal current stress.

On n+k portion (fig. 1) polarity current stress change on opposite and the MOS-structure recharge begins. On beginning portion (fig. 1) where through MOS-structure injection and capacitive currents simultaneously proceed, dividing them by a technique [6] and having taken advantage of the equation (4), having replaced I_3 on I_{n+k}, also it is possible to calculate *I-V* the characteristic in a range of currents more low I_{n+k}. Then on n+k portion the injection current becomes equal to zero (fig. 1, c), and in this case n+k portion on dependence $V_I(t)$ is equivalent to 2-th portion. Then, using equations (1, 3, 5) and having replaced in them I_2 on I_{n+k}, on n+k portion, also it is possible to calculate low-frequency *C-V* dependence and to define MOS-structure parameters right after high-fields injections (Fig. 1, portion B).

Such way of reception *C-V* and *I-V* characteristics, within the limits of the same technique, allows to determine characteristics of charging degradation right after tunnel injection, having lowered influence relaxation processes. The subsequent realization of the offered algorithm at opposite polarity applied current allows to determine density charge, capture cross-sections of charging traps and centroid a charge in dielectric films.

Having received experimental the current-voltage characteristic, we choose the corresponding theoretical model of electron transport. Thus, for thermal SiO_2 films the transport under high-fields, as a rule, by the Fowler-Nordheim model.

Having received experimental current-voltage characteristic, it is possible to define such important electrophysical characteristic MOS-structure, as value of a potential barrier on injecting interface φ_B and effective mass of electron m* in film SiO_2. On the other hand, at use of the given technique in the conditions of manufacture when the test is carried out on the MOS-structures made on one technology, and one of parameters φ_B or m^*, or both together, the control of a thickness of oxide d_{ox} which is the important technological parameter is a priori known, possible. On the found values of capacity of MOS-structure, with the account of its geometrical sizes, dielectric constant of oxide is determined.

Nowadays, the constant current stress technique [1, 5] is widely used for investigation of positive charge generation in MOS structures. In this technique, a constant current pulse is applied to a MOS structure, causing the charging of structure capacitance and initiating the mode of Fowler-Nordheim high-field tunnel electron injection. Measuring the MOS structure voltage shift one can estimate the charge state change of a sample [1, 5].

However, under high densities of tunnel current, the positive charge generation rate increases. Hence, when an injection current begins to flow in a gate dielectric, it can generate a part of positive charge, not taking into account in this technique. As a result, the constant current technique at high densities of injection current can give underestimated values of positive charge. The usage of other techniques of positive charge value determining, for example C-V technique or charge pumping, requires the interruption of injection and the re-commutation sample. It can lead to relaxation of a part of the positive charge and significant errors in its determination.

To eliminate this shortcoming, a special algorithm to control the current stress applied to a sample was proposed, shown on Fig. 2(a). Two levels of current are applied to the structure: stress level I_S and measuring level I_m. Positive charge generation in the gate dielectric is caused by stress current value I_S. The current I_m must have a value of several orders less than I_S, to minimize the dielectric charge degradation by this current. The MOS charge state change can be estimated by the voltage shift on MOS structure at current I_m (Fig. 2,b). Voltage shift under current I_m takes into account positive charge generated in the gate dielectric right after the switching to the current I_S. Thus, the inherent errors of constant current technique are reduced when using our technique.

For investigation of positive charge generation, the pulse short-time change of current level from Is to I_m and from I_m to I_S can be used (Fig. 2). The stress mode duration t_s depends on the monitoring discreteness of positive charge, it can be varied on different stages of injection. The measuring mode duration tm must provide the establishing of injection mode, but it should not affect significantly on the sample charge state.

The approbation of present technique was carried out during the investigation of bulk positive charge relaxation, generated by high-field tunnel injection of electrons in thermal film of SiO_2. It was found out that in injection fields, the relaxation rate of bulk positive charge in SiO_2 film increased with the rise of electric field.

Algorithm of change in stress current in the study of relaxation of the positive charge is shown in Fig. 2,c (curve 1). Initially constant current pulse amplitude I_m is applied to the MOS structure, at which there is a charge capacity of MOS structure, and then occurs in the mode of tunnel on Fowler-Nordheim injection of electrons in the dielectric. The amplitude of current I_m is chosen from the condition that the duration of the stress current that is applied to the MOS structure for the time necessary for the study of relaxation processes have not led to noticeable changes in the charge state of the sample. In Fig. 2,d (curve 1) shows the high-voltage portion of voltage changes on MOS structure in the study of relaxation of the positive charge of the proposed method. After injection into the dielectric required value of the charge is a stepped decrease in the amplitude of a current of up to the value of I_m (Fig. 2,c curve 1). Then, maintaining current constant at the level of I_m, for changes of the voltage on the MOS structure is monitored by the relaxation of the positive charge. Study of the processes of relaxation of a charge in the dielectric under currents, other than I_m or in noninjection mode, realized by applying to the MOS structures by injection current I_a or voltage V_a, respectively. The change of the charge state of the structure in this case is measured by short-time switching current stress in the mode of injection, corresponded the current I_m, and is measured the V_{Im} (Fig.2).

As part of the proposed technique can be carried out statistical measurement insulation defects and defects of the charge stability of the dielectric layers of MOS structures on the wafer. Test of defects in insulation is based on the measurement of set voltage leakage (voltage - which flows through the dielectric given current density). Monitor of the charge stability defects of gate dielectric MOS structures is based on the measurement of the charge injected into dielectric to breakdown. Moreover, the set leakage voltage and the charge injected to breakdown are measured sequentially in the present technique, without switching the sample.

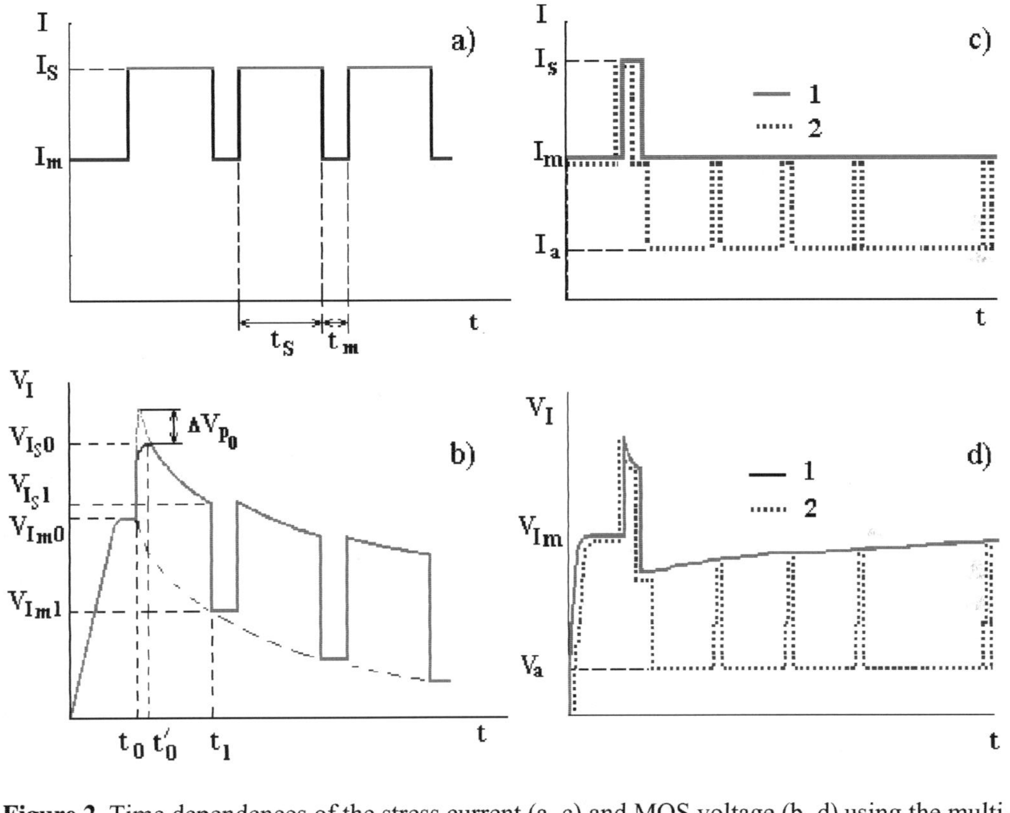

Figure 2. Time dependences of the stress current (a, c) and MOS voltage (b, d) using the multi-level current stress technique to study the generation of (a, b) and relaxation (c and d) the positive charge in an injection and no injection modes. In each group, curves 1 and 2 refer to two- and three-level injection, respectively.

Figure 3 shows an algorithm for measuring the set leakage voltage and the charge injected to breakdown in this technique. On sections I and II, figure 3 is measured by the set leakage voltage, then the results of its measurement is constructed corresponding to the histogram of MOS structures. Structures that have no insulation defects get to the main peak of the histogram distribution for set leakage voltage, after that for these structures measured the charge injected to breakdown (sections 3 and 4 in figure 3). In fig. 3 shows the histogram measured on MOS-capacitors located in the test modules of CMOS IC. The ICs were fabricated on phosphorus-doped (resistivity 4.5 Ω×cm) n-type Si <100> wafers. 25 nm-thick silicon dioxide was thermally grown at 850°C. The polysilicon film doped with phosphorus was used as a gate.

To apply the proposed method of test of defects in the dielectric layers of MOS structures, implementing a unified approach to the study and monitor of insulation and the charge stability defects of required to implement the following procedure for the measurement and testing:

- Selected by measuring levels of stress and the injection current, based on the area of MOS structures, the thickness of the dielectric, time limits;
- Measured time dependence of the voltage on the MOS structure for measuring and stress injection currents;
- Determined by measurement of the time until t_{BD};
- Carried out measurements of the parameters for the MOS structures;
- Building the combined histograms of MOS structures for set leakage voltage and the charge injected to breakdown figure 3 (c);
- Determined by the quantitative characteristics of groups of MOS structures.

Using the proposed technique allowed to choose modes of plasma treatment of MOS structures, for which the density decreases and the rate of generation of new electron traps in the thin film of SiO_2, resulting from the injection of electrons in high-fields. This can significantly improve the charge stability of the structure and to increase such an important parameter as the value of the charge, which can be injected into the dielectric to breakdown of the sample [6].

3. Conclusion

The injection technique for testing and studying the thin gate dielectric of MOS structures has been proposed. This technique allows to:

- measure change in the charge state of the MOS structures and to produce charge injection into dielectric using the same technique;
- evaluate the effective charge of dielectric, the interface state density, the flat band voltage, the surface potential value for a given bias, the potential barrier height on the injection interface, the dielectric capacitance;
- determine the variations of the voltage depended on the injected charge, the concentration of the charge traps and their capture cross-sections, the value of the charge injected into dielectric until the sample breakdown occurs as well as to locate the charge centroid in dielectric.
- investigate the generation and relaxation processes in thin gate dielectric of MOS structures during and after high-field stress.

The new injection technique for quality control of dielectric layers of MOS structures based on an analysis of the time dependence of voltage on the structure when applying for her constant current pulses. An algorithm for charge injection into the dielectric and the algorithm measuring characteristics in the system parameters of quality control gate dielectric MOS devices. We propose a visual presentation of the results of test isolate defects and defects of the charge stability in the form of histograms of the distributions of the combined MOS structures for set leakage voltage (V_{sl}) and charge injected to breakdown (Q_{inj}^{BD}).

Acknowledgments

The work was supported by grant No. 12-02-97533 of RFBR and Kaluga's region and the Federal Target Program «Scientific and pedagogical staff of innovative Russia» on 2009 - 2013 (the State contract No. 16.740.11.0151).

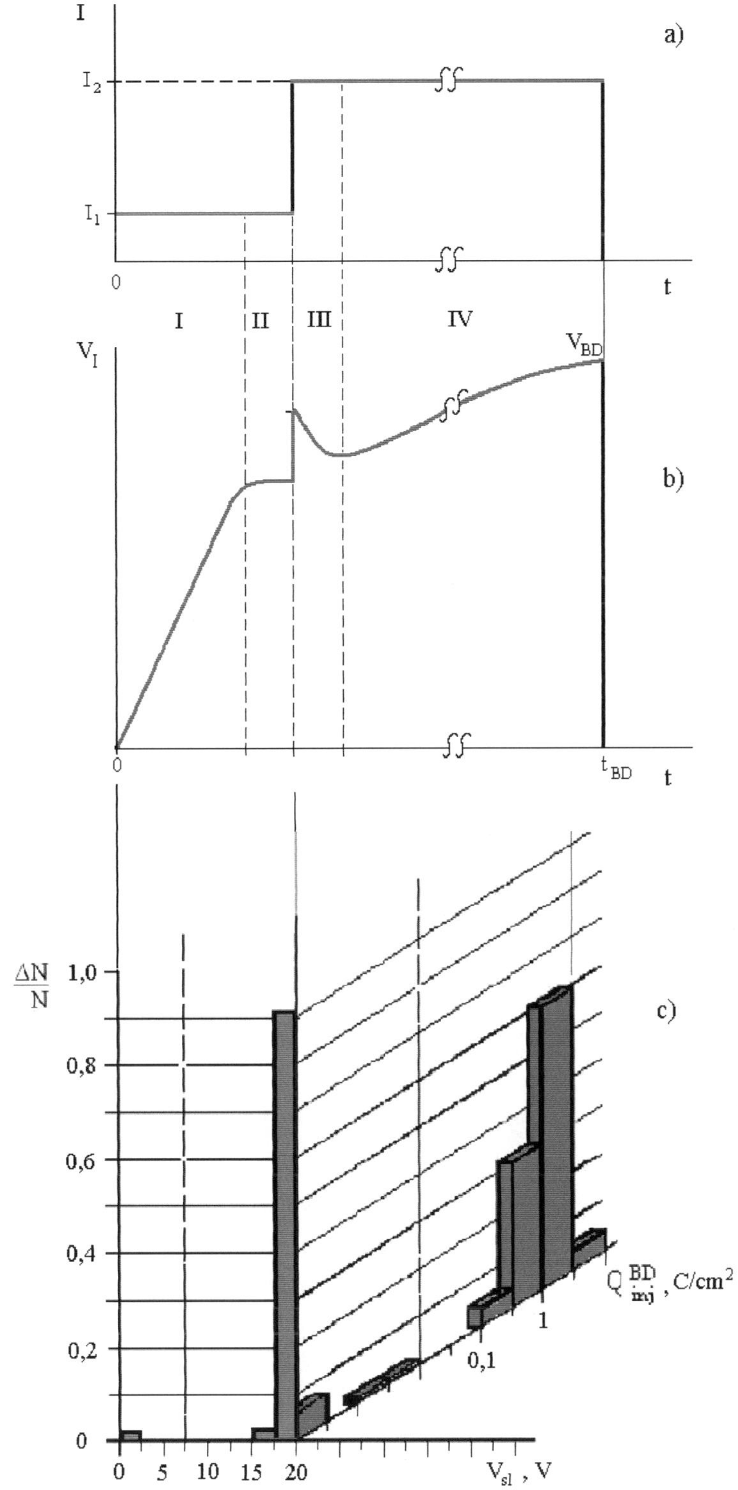

Figure 3. Time dependences of the stress current (a) and MOS voltage (b), as well as the combined histogram MOS structures set leakage voltage (V_{sl}) and charge, injected to breakdown (c).

IOP Conf. Series: Materials Science and Engineering **41** (2012) 012017 doi:10.1088/1757-899X/41/1/012017

References

[1] Lombardo S, Stathis J, Linder P, Pey K, Palumbo F, Tung C 2005 *J. Appl. Phys.* **98** 121301
[2] Green M, Gusev E, Degraeve R, Garfunkel E 2001 *J. Appl. Phys.* **90** 2057
[3] Sze S.M., Ng K.K. 2007 *Physics of Semiconductor Devices* (Inc., Hoboken, New Jersey)
[4] Arnold D, Cartier E, DiMaria D 1994 *Phys. Rev. B* **49** 10278
[5] DiMaria D, Cartier E, Buchanan D 1996 *J. Appl. Phys.* **80** 304
[6] Bondarenko G, Andreev V, Loskutov S, Stolyarov A 1999 *Surface and Interface Analysis* **28** 142
[7] Bondarenko G, Andreev V, Stolyarov A, Tkachenko A 2002 *Vacuum* **67/3–4** 507
[8] Bondarenko G, Andreev V, Maslovsky V, Stolyarov A, Drach V 2003 *Thin solid films* **427** 377
[9] Andreev V, Baryshev V, Bondarenko G, Stolyarov A 2003 *Russian Microelectronics* **32** 119
[10] Bondarenko G, Andreev V, Drach V, Loskutov S, Stolyarov M 2006 *Thin solid films* **515** 670

IOP Conf. Series: Materials Science and Engineering **41** (2012) 012018 doi:10.1088/1757-899X/41/1/012018

Resistive switching of Ti/HfO_2-based memory devices: impact of the atmosphere and the oxygen partial pressure

T Bertaud[1,3], M Sowinska[1], D Walczyk[1], Ch Walczyk[1], S Kubotsch[1], Ch Wenger[1] and T Schroeder[1,2]
[1] IHP, Im Technologiepark 25, 15236 Frankfurt Oder, Germany
[2] Brandenburgische Technische Universität, Konrad-Zuse-Strasse 1, 03046 Cottbus, Germany

E-mail: bertaud@ihp-microelectronics.com

Abstract. The electrical characteristics of different resistance states (virgin, OFF and ON) of a Ti/HfO_2/TiN metal-insulator-metal device for resistance random access memory are investigated under different gas ambient. The influence of the atmosphere, the total pressure and the oxygen concentration during electrical measurements is underlined thanks to retention (*I*-t) and impedance spectroscopy (*Z*-f) measurements. The total pressure influences the current levels of the three different resistive states: when the total pressure decreases, the current increases, probably due to an increase of the concentration of oxygen vacancies in the HfO_2.

1. Introduction

Ti/HfO_2-based Resistance change Random Access Memory (RRAM) is a very promising candidate for embedded nonvolatile memory because of compatibility with silicon-based complementary metal-oxide-semiconductor technologies, low-power dissipation and memory block integration [1-4]. However, there are still many challenges which limit the RRAM technology development, in particular reliability and a detailed understanding of the resistive switching (RS) mechanism. Typically, two pictures of the RS mechanism are considered, both involving charged oxygen vacancies ($V_O^{\cdot\cdot}$) and locating the switching event at the metal-dielectric interface. The local growth and rupture of $V_O^{\cdot\cdot}$-based nanofilaments between the bottom and top electrode [5] or the movement of these $V_O^{\cdot\cdot}$ at the metal/oxide interface induced by an electric field [6]. In previous hard X-ray photoelectron spectroscopy (HAXPES) studies, we underlined the role of $V_O^{\cdot\cdot}$ in the electroforming process of the RS [7] and proposed a model for the RS mechanism [8] for Ti/HfO_2/TiN-based system. As previously reported in literature, the ambient atmosphere has also shown an impact on the electrical characteristics of MIM devices [9-10]. In this contribution, the impact of the atmosphere and the oxygen concentration during measurements on the electrical characteristics of different resistance states (virgin, OFF and ON) of Ti/HfO_2/TiN MIM devices is studied.

2. Experimental details

The preparation of the Ti-10 nm/HfO_2-14 nm/TiN-112 nm metal-insulator-metal (MIM) structure can be found elsewhere [3]. The TiN top capping layer was not deposited in order to better analyse the impact of the gas ambient on the electrical characteristics of the MIM structure. In order to avoid the

[3] To whom any correspondence should be addressed.

contacting issues on the 10 nm thick top electrode, a special setup was realized as follows. A sample with a size of 1×1 mm^2 was prepared with the bottom electrode contact (200×200 μm^2 in area) opened by time of flight secondary ion mass spectrometry (ToF-SIMS) depth profiling. As shown in figure 1, the sample was then mounted on a printed circuit board (PCB) and wire-bonded to external electrical connectors. Electrical characterizations (current voltage, retention and impedance spectroscopy) of RRAM devices were performed using a Keithley 4200 SCS semiconductor characterization system. All DC voltages were applied to the top electrode, whereas the bottom electrode was electrically grounded. All these electrical measurements were performed in a vacuum chamber which is coupled with O_2/N_2 gas providers and an oxygen sensor allowing measurements under different gas ambient. The electrical characteristics of different resistive states (virgin, OFF and ON) of the MIM diodes were studied under standard atmosphere (pressure p = 1 bar), vacuum ($p \sim 3\times10^{-5}$ mbar) and oxygen atmosphere (p ~ 12 mbar). The measurements were realized at room temperature (RT).

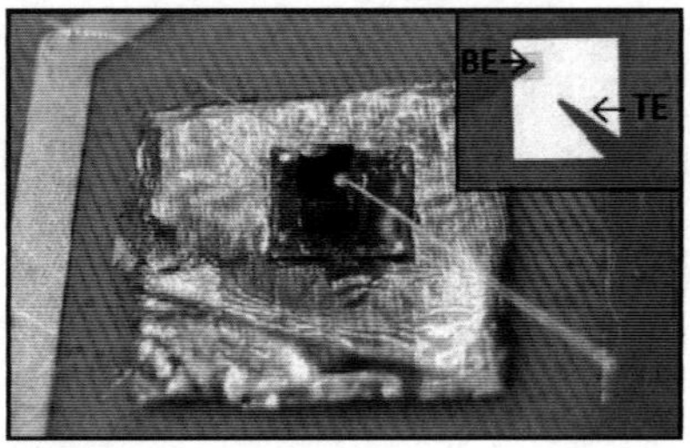

Figure 1. Wire-bonded 1×1 mm^2 MIM device mounted on a PCB. The inset shows a top view of the sample with the bottom contact opened by ToF-SIMS.

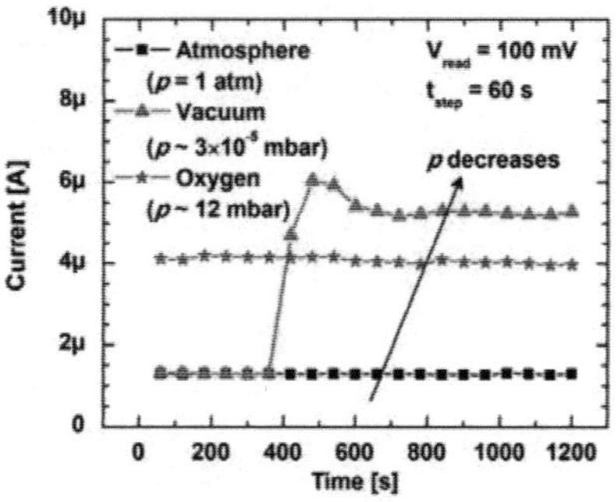

Figure 2. Retention characteristics of the fresh MIM device under the three different atmospheres. The current level was read every minute at 100 mV for 20 minutes.

3. Characterization of the virgin-state

First, the virgin-state was studied using retention and impedance spectroscopy measurements. Figure 2 presents the retention characteristics. Reading is performed at 100 mV every minute, under three studied atmospheres. From the characteristics under vacuum, we observe that the system needs a certain time to stabilize and reach equilibrium with the ambient atmosphere (the measurements under the standard and oxygen-rich atmospheres were performed after waiting approximately 1 hour). After that, it is noticed that the total pressure p impacts the current levels: when the total pressure decreases, the current increases. This can be linked to an increase of the concentration of $V_O^{\cdot\cdot}$ in the HfO_2. It is known that $V_O^{\cdot\cdot}$ are stable in the form of double ionized species and thus act as n-type donor [11].

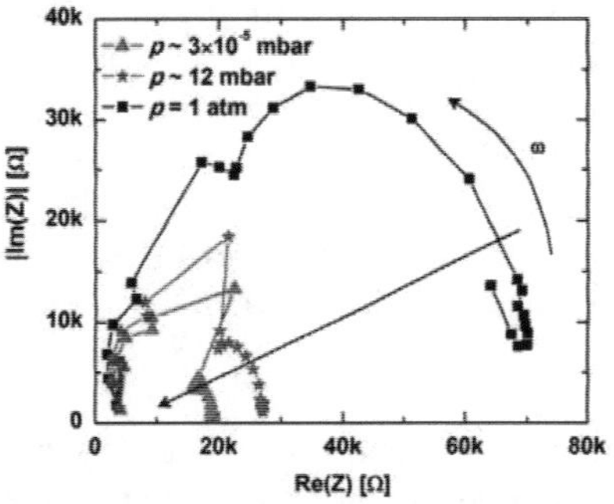

Figure 3. Cole-Cole plots from impedance (Z) spectroscopy by applying 100 mV AC small signals (10 kHz – 1 MHz).

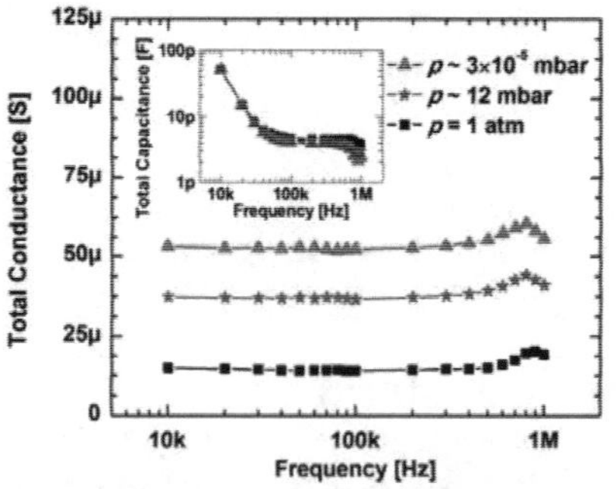

Figure 4. Total conductance and capacitance (in the inset) obtained from the impedance spectroscopy measurements.

To further analyze this impact of the total pressure, we performed impedance spectroscopy measurements. Figure 3 presents the Cole-Cole plots of the MIM device under three different atmospheres. As expected for MIM capacitor devices, semi-circle plots are obtained and the total pressure has a strong impact on these characteristics. In a first approximation, the quasi semi-circle suggests that the device can be modelled by a parallel resistor-capacitor circuit [12]. We thus modelled the impedance Z with a parallel conductance and capacitance and present in figure 4 their evolution with the frequency. We clearly observe an increase of the total conductance with the decrease of the total pressure whereas no clear impact is observed on the total capacitance (in the inset). This result suggests that the observed increase of the current with the decrease of the pressure is linked to an increase of the conductance. This is probably due to an increase of the $V_O^{\cdot\cdot}$ concentration at the Ti/HfO_2 interface which lowers the barrier height [7].

4. Characterization of the OFF- and ON-states

After an electroforming process, the MIM device presents bipolar RS behavior at all atmospheres, as shown in figure 5. A change of the switching polarity was observed between the standard atmosphere and the low-pressure measurements. A positive set process is obtained for the electrical measurements done under the standard atmosphere whereas a negative set voltage is found for the measurements at 3×10^{-5} mbar (vacuum) and 12 mbar (oxygen atmosphere). The change of switching polarity was reported in literature for $SrTiO_4$ thin films and linked to a saturation of the $V_O^{\cdot\cdot}$ concentration at the interface [13]. However, this behavior is still under scientific debate for HfO_2-based RRAM.

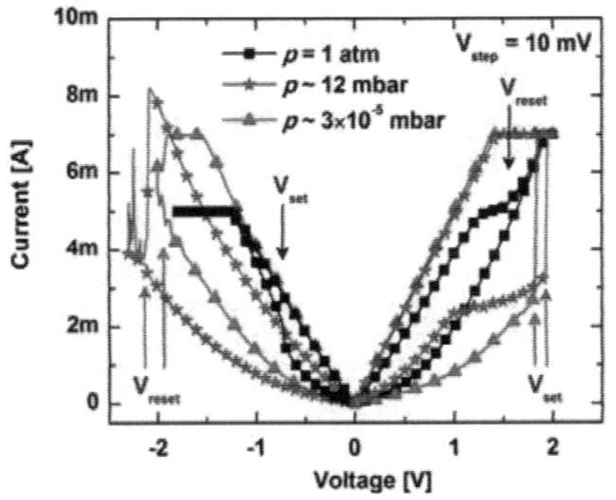

Figure 5. RS curves obtained at the three different atmospheres.

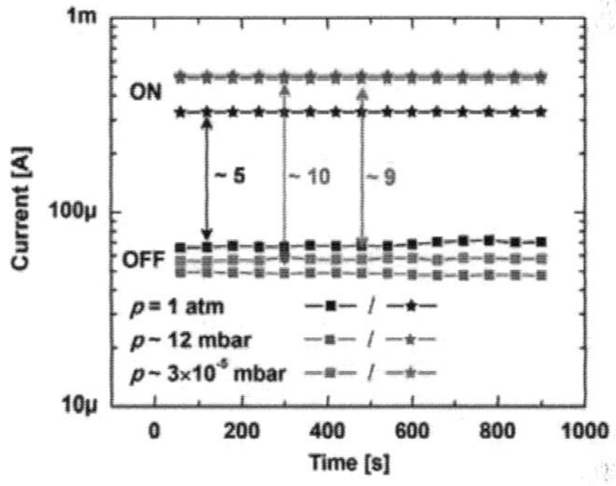

Figure 6. Retention measurements for the OFF- and ON-states.

In order to further analyse the characteristics of the OFF- and ON-states, we performed retention and impedance spectroscopy measurements. In figure 6, one observes that the total pressure impacts the R_{OFF}/R_{ON} ratio: a bigger ratio was found at the lowest pressures compared to the standard atmosphere. This can be linked to the work of Chen *et al.* which have shown the possibility to further improve the RS characteristics by optimizing the encapsulation of the RRAM device with an oxygen-poor layer [14]. Then, we present in figure 7 the Cole-Cole plot of the impedance spectroscopy measurements done for the OFF- and ON-states at the three different atmospheres. Compared to the virgin-state, the data points of the OFF- and ON-states cluster near the Re(Z) axis and suggest that the memory cell is shunted by a conductance [12]. Besides, no clear differences are visible between the three different atmospheres (see the inset). To better distinguish the differences, we extracted in figure 8 the total conductance of the device versus frequency. For the OFF- and ON-states, the frequency behavior of the conductance is identical; a very constant response is observed over the studied frequency range (from 10 kHz to 1 MHz). This indicates the presence of the same conduction mechanism for the different atmospheres. However, quantitative differences occur: minor change in the OFF-state and a clear conductance increase with the pressure decrease for the ON-state. Given the limited diffusion length at low temperatures, only the Ti/HfO_2 interface will see an evolution of the $V_O^{\cdot\cdot}$ concentration with the pressure. According to the switching physics derived from HAXPES studies on the

IOP Conf. Series: Materials Science and Engineering **41** (2012) 012018 doi:10.1088/1757-899X/41/1/012018

Ti/HfO_2/TiN system, we can speculate that the ON-state is more dependent than the OFF-state on the pressure because of higher band bending at this interface (i.e. higher density of $V_O^{\cdot\cdot}$) [8]. Further studies need to be performed to better understand the role of $V_O^{\cdot\cdot}$, for example by coupling temperature measurements at different atmospheres in order to thermally activate the diffusion phenomenon.

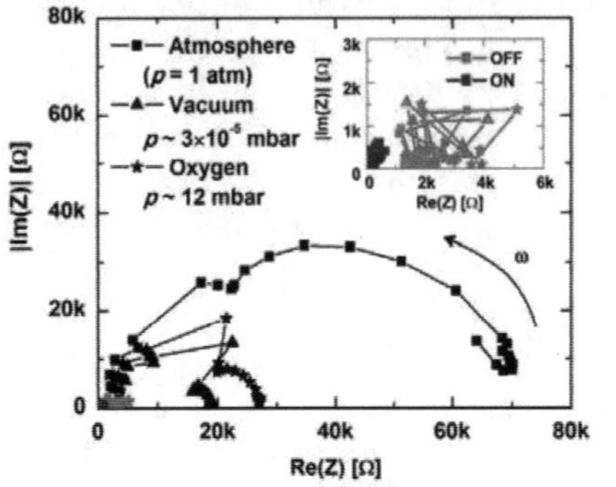

Figure 7. Cole-Cole plots of the OFF- and ON-states (in the inset) compared with the virgin-state.

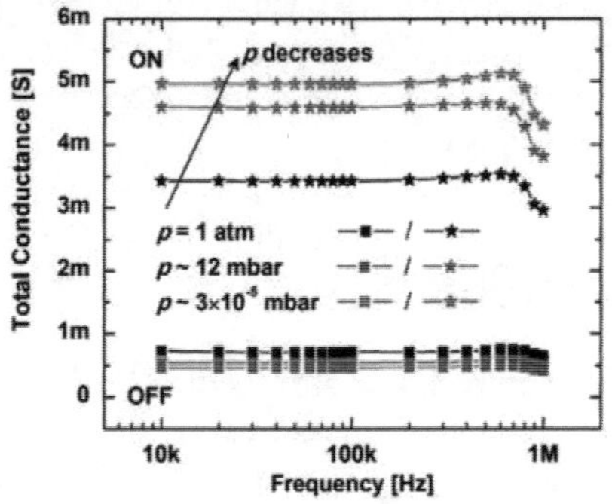

Figure 8. Total conductance of the device in the OFF- and ON-states under different atmospheres.

5. Conclusions

The impact of the atmosphere on the electrical properties of Ti/HfO_2/TiN-based MIM structures was studied at different gas ambient: standard atmosphere (p = 1 bar), vacuum ($p \sim 3\times10^{-5}$ mbar) and oxygen atmosphere (p ~ 12 mbar). The measurements, done at RT for three different resistive states (virgin, OFF and ON), indicate an increase of the conductance and the current level with the decrease of the total pressure p, with a higher influence on the virgin- and ON-states. This observation underlines the role of $V_O^{\cdot\cdot}$ in the RS properties of a RRAM MIM stack, as also pointed out in our HAXPES studies [7-8]. In addition, the varying R_{OFF}/R_{ON} ratio shows the possibility to improve the RS parameters by engineering the encapsulation of the RRAM device with an oxygen-poor film [14].

6. References

[1] Sheu S-S *et al.* 2011 Proc. IEEE International Solid-State Circuits Conference (ISSCC) **200**
[2] Walczyk D *et al.* 2011 *Microelectron. Eng.* **88** 1133
[3] Walczyk Ch *et al.* 2011 *IEEE Trans. Electron Devices* **58** 3124
[4] Bertaud T *et al.* 2012 *Thin Solid Films* **520** 4551
[5] Yang J J *et al.* 2009 *Nanotechnology* **20** 215201
[6] Akinaga H and Shima H 2010 Proc. of the IEEE **98** 12 2237
[7] Sowinska M *et al.* 2012 *Appl. Phys. Lett.* **100** 233509
[8] Bertaud T *et al.* 2012 *Appl. Phys. Lett.* submitted
[9] Guha S and Narayanan V 2007 *Phys. Rev. Lett.* **98** 196101
[10] Goux L *et al.* 2010 *Appl. Phys. Lett.* **97** 243509
[11] Robertson J, Sharia O and Demkov A A 2007 *Appl. Phys. Lett.* **91** 132912
[12] Yu S, Jeyasingh R, Wu Y and Philip Wong H S 2011 *Appl. Phys. Lett.* **99** 232105
[13] Shibuya K, Dittmann R, Mi S and Waser R 2010 *Adv. Mater.* **22** 411
[14] Chen Y S et al. 2011 *IEEE Electron Dev. Lett.* **32** (3) 390

Acknowledgments

The authors thank H Silz, D Kot, D Martynenko, M Elkhouly and F Popiela from IHP for their help in sample preparation. The authors gratefully acknowledge financial support from the Deutsche Forschungsgemeinschaft (DFG) for the RRAM project under contract SCHR 1123/7-1.

IOP Conf. Series: Materials Science and Engineering **41** (2012) 012019 doi:10.1088/1757-899X/41/1/012019

Internal Photoemission at Interfaces of ALD $TaSiO_x$ Insulating Layers Deposited on Si, InP and $In_{0.53}Ga_{0.47}As$

H Y Chou[1], V V Afanas'ev[1], N H Thoan[1 3], C Adelmann[2], H C Lin[2], M Houssa[1] and A Stesmans[1]

[1]Department of Physics, University of Leuven, Belgium

[2]IMEC, Leuven, Belgium

[3]Insitute of Engineering Physics, Hanoi University of Technology, Hanoi, Vietnam

HsingYi.Chou@fys.kuleuven.be

Abstract. Electrical analysis of interfaces of (100)Si, (100)InP, and (100)$In_{0.53}Ga_{0.47}As$ with $TaSiO_x$ (Ta/Si≈1) films atomic-layer deposited using $SiCl_4$, $TaCl_5$, and H_2O precursors suggests Ta silicate as a good insulating and surface passivating layer on all three semiconductors. However, when a positive voltage is applied to the top metal electrode in a metal/ $TaSiO_x$ /semiconductor configuration, considerable hysteresis of the capacitance-voltage curves, both at 300 and 77 K, is universally observed indicating electron injection and trapping in the insulator. To shed some light on the origin of this charge instability, we analyzed interface band alignment of the studied interfaces using the spectroscopies of internal photoemission and photoconductivity measurements. The latter reveals that independently of the semiconductor substrate material, $TaSiO_x$ layers exhibit a bandgap of only 4.5±0.1 eV, typical for a Ta_2O_5 network. The density of electron states associated with this narrow-gap network may account for the enhanced electron injection and trapping. Furthermore, while a sufficiently high energy barrier for electrons between Si and $TaSiO_x$ (3.1±0.1 eV) is found, much lower IPE thresholds are encountered at the (100)InP/$TaSiO_x$ and (100) $In_{0.53}Ga_{0.47}As/TaSiO_x$ interfaces, i.e., 2.4 and 2.0 eV, respectively. The lower barrier may be related by the formation of narrow-gap In-rich interlayers between $A_{III}B_V$ semiconductors and $TaSiO_x$.

1. Introduction

Atomic-layer deposited (ALD) tantalum silicate ($TaSiO_x$) has recently been suggested as a viable gate insulator for the use in III-V MOSFET devices [1]. This insulating oxide delivers a reasonably low density of interface traps on $A_{III}B_V$ semiconductors, such as InP and $In_{0.53}Ga_{0.47}As$, a dielectric permittivity higher than that of Al_2O_3 (κ=11 for Ta:Si= 1:1, $\kappa\approx$ 15 for Ta-rich material)[2] and, furthermore, remains amorphous after thermal processing. With these superior properties, it is naturally to extend analysis to the electronic structure of ALD $TaSiO_x$ and, further, to the band alignment at its interfaces with different semiconductors. Here, by using internal photoemission (IPE) and photoconductivity (PC) measurements we determined the electron band alignment between Si, InP and $In_{0.53}Ga_{0.47}As$ with ALD $TaSiO_x$. In particular, it is found that oxidation of the $A_{III}B_V$ semiconductors surface results in formation of an In-rich interlayer (IL) with a narrow gap, leading to low interface barriers for electrons and causing charge instability of the insulator.

IOP Conf. Series: Materials Science and Engineering **41** (2012) 012019 doi:10.1088/1757-899X/41/1/012019

2. Experimental

The silicate layers $TaSiO_x$ (Si: Ta≈ 1) were deposited by ALD using $TaCl_5$, $SiCl_4$ and H_2O at 250 °C onto (100)-oriented surfaces of Si, InP or $In_{0.53}Ga_{0.47}As$. Four target thicknesses of the $TaSiO_x$ layer (5, 10, 15 and 20 nm) were achieved by altering the number of cycles of different precursors and the oxidizer. For IPE and PC study, MOS capacitors were fabricated by evaporation of semitransparent (13-nm thick) Au electrodes of 0.4 mm^2 area on top of $TaSiO_x$ layer. The photocurrent was measured as a function of photon energy ($h\nu$) in the range from 1.8 to 5.5 eV with the applied metal gate bias ranging from -2.0 to +2.5 V. The quantum yield (Y) was defined by normalizing the measured photocurrent to the incident light flux. Further, it was used to determine spectral thresholds of photocurrent which provide the information regarding the bandgap width and the band alignment.

3. Results and Discussion

Since silicon represents a well studied substrate crystal we first used the Si-based MOS structures to evaluate the bandgap width of ALD $TaSiO_x$ and band alignment. The spectra of quantum yield measured on a n-Si/$TaSiO_x$ (20nm)/Au sample with different gate biases are shown in Fig. 1(a). The photocurrent observed under positive bias (open symbols) is associated with IPE of electrons from the Si substrate into $TaSiO_x$. This could be confirmed by the observation of a field independent optical singularity [E_1(Si)] at $h\nu$ = 3.4 eV in Fig. 1(a) corresponding to excitation of direct optical transitions in the Brillouin zone of Si [3,4]. The spectral threshold was determined by linear extrapolating the $Y^{1/3}$- $h\nu$ plot (cf. the insert) to zero yield after subtracting the sub-threshold photocurrent probably caused by excitation of defect states. This inferred spectral threshold has a value Φ_e(Si)= 3.1±0.1 eV, weakly depending on the strength of electric field and corresponding to the energy barrier between the top of the Si valence band (VB) and the bottom of the $TaSiO_x$ conduction band (CB).

Next, the bandgap of the $TaSiO_x$ layer was found using the PC yield plotted in $Y^{1/2}$-$h\nu$ coordinates (Fig.2 (b)) to enable extraction of the threshold. The same spectral threshold was found for both orientations of the electric field in the oxide suggesting its relationship to the optical generation of electron-hole pairs in the bulk of the oxide, thus yielding the bandgap of ALD-$TaSiO_x$ of $E_g(TaSiO_x)$= 4.4±0.1 eV. Within the experimental accuracy of 0.1 eV, this value coincides with the bandgap of amorphous Ta-oxide reported elsewhere [5], suggesting that in the Ta silicate the Ta-oxide sub-network preserves the same gap value as in the elemental metal oxide.

If considering the inferred bandgap width and the IPE threshold together, one arrives to a low VB offset, only 1.3 eV, between Si and $TaSiO_x$ which is inconsistent with the observed leakage currents below 10^{-11} A/cm^2 at 1 MV/cm field in $TaSiO_x$. Then the IPE of electrons from Si probably occurs into the CB of the Si-oxide sub-network in the $TaSiO_x$. It is really surprising that with the revealed 4.4 eV-wide bandgap of ALD-$TaSiO_x$, the Si-oxide sub-network still holds sufficiently high barriers to suppress the leakage current.

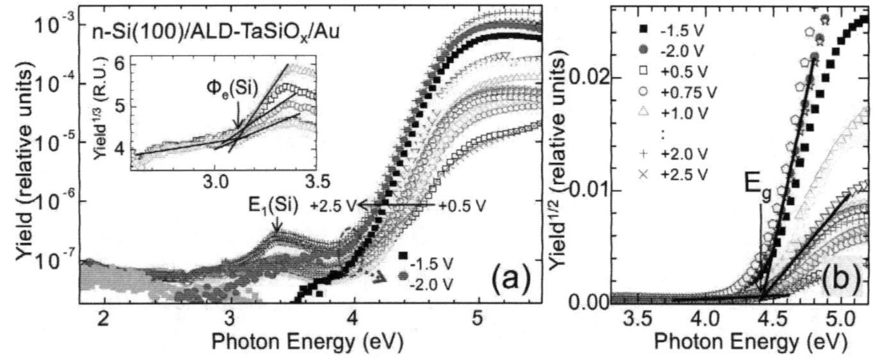

Fig. 1. IPE/PC yield (a) and its square root (b) as a function of photon energy measured in Si/TaSiOx/Au samples with the applied bias varying from +0.5 to 2.5 V (open symbols) and from -1.5 to -2.0 V (f lled symbols).

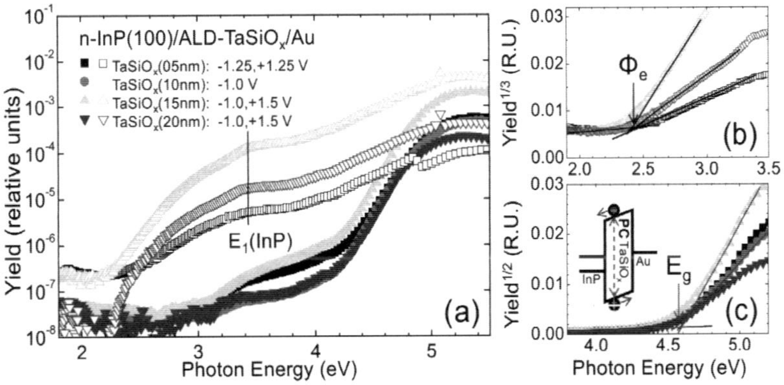

Fig. 2. IPE/PC yield (a) and cube (b) and square root (c) of it as a function of photon energy measured in InP/TaSiOx/Au samples with different insulator thickness. The applied biases are indicated in the legend with open and f lled symbols, corresponding to positive and negative voltages, respectively.

The IPE/PC yield spectra observed on (100)InP MOS samples with different thickness of $TaSiO_x$ are compared in Fig. 2. The semi-logarithmic plot of the quantum yield under different positive biases (panel (a)) exhibits onset at photon energy of ~2.4 eV, but then ceases to increase at $h\nu$= 3.3eV, which coincides in energy with the critical point E_1 of InP [6]. This field independent critical point corresponds to the excitation of a direct optical transition from $L_{4,5}^{V}$ to L_6^{C} high-symmetry points in the Brillouin zone of InP thus revealing the latter as the source of photoelectrons. After establishing that the photocurrent is related to electron IPE from InP, the spectral threshold corresponding to the barrier height between the VB of InP and the CB bottom of $TaSiO_x$ can be found as Φ_e(InP)= 2.4±0.1 eV using the $Y^{1/3}$- $h\nu$ plot as shown in Fig. 2(b). As no signatures of the InP electronic structure are observed in the yield spectra taken under negative bias (filled symbols in Fig. 2(a, c)), a different excitation mechanism must be involve to explain the photocurrent generation. Similar to the earlier analyzed Si MOS samples, the onset of a photo current at E_g= 4.6±0.1 eV can be associated with the optical excitation of electrons across the bandgap in the amorphous ALD- $TaSiO_x$ layer which is consistent with the result obtained in Si/ $TaSiO_x$ /Au capacitors.

Contrary to the cases of Si and InP, the electrons from $In_{0.53}Ga_{0.47}As$ are excited into the oxide at photon energies above 2 eV (positive bias) with spectral curves bearing no signatures of optical transitions within the substrate crystal [open symbols in Fig. 3(a)]. This observation excludes IPE from $In_{0.53}Ga_{0.47}As$ as a possible source of the photocurrent in the spectral range 2< $h\nu$ <4 eV, i.e., it is likely that this signal is generated through optical excitation of electron states at the interface which are not related to the semiconductor electron bands. Rather, optical excitation of an In-rich IL which presence has been revealed for $TaSiO_x$ grown both on InP and $In_{0.53}Ga_{0.47}As$ by transmission electron microscopy and secondary ion mass spectroscopy experiments (not shown here), is the most probable

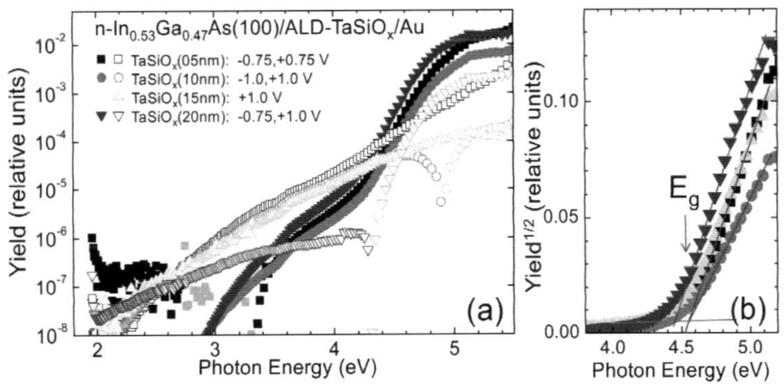

Fig. 3. IPE/PC yield (a) and square root (b) of it as a function of photon energy measured in $In_{0.53}Ga_{0.47}As/TaSiO_x/Au$ samples with different thickness of the insulator. The applied biases are as indicated in the legend with open and f lled symbols, corresponding to positive and negative voltages, respectively.

source of the observed photocurrent since In_2O_3 is known to poses a narrow bandgap (2.7-2.9 eV) [9] being semiconducting rather than insulating material. The bandgap of ALD grown $TaSiO_x$ could also be inferred from the PC spectra shown in Fig. 3(b) yielding, E_g= 4.5 ±0.1 eV, i.e., close to that found on the other substrates (Si and InP). As a final step we compared the spectra of the IPE quantum yield from three studied semiconductors into Al_2O_3 or $TaSiO_x$, as measured under positive bias on the top Au electrode in Fig. 4(a) and 4(b), respectively. The threshold IPE from the Si VB into the CBs of Al_2O_3 [5] and $TaSiO_x$ are found to be 3.2 and 3.1 eV, respectively. However, the threshold trend observed for Al_2O_3 and $TaSiO_x$ on InP and $In_{0.53}Ga_{0.47}As$ appears to be very different, with photocurrent thresholds shifted towards lower photon energy in the last case [7, 8]. This discrepancy can be associated with oxidation of the In-containing semiconductor during ALD process leading to formation of an In-rich interlayer. By comparing the results from the $In_{0.53}Ga_{0.47}As$ and InP samples as shown in Fig.4 (a) and (b), the photocurrent threshold lowering is found to be around 1.4 and 1.6 eV for $In_{0.53}Ga_{0.47}As$ and InP interfaces, respectively. Again, this barrier lowering could be explained by the presence of a interlayer with small bandgap, probably indium oxide which $E_g(In_2O_3)$ has been estimated to be 2.7-2.9 eV [9]. Apparently, the absence of Trimethylaluminium during ALD of TaSiOx prevents removal of In oxides (no self-cleaning) thus impairing the interface properties in terms of barrier height.

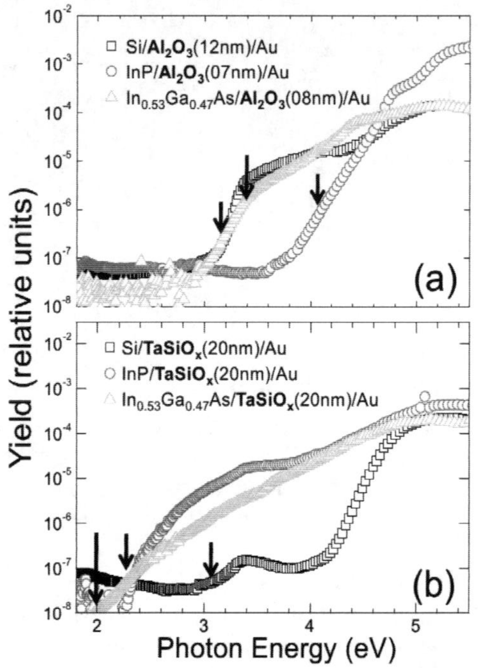

Fig. 4. Comparison of the IPE yield spectra for several semiconductors with Al_2O_3 (a) and $TaSiO_x$ (b) insulating layers on top, measured under positive bias applied to the top metal electrode. The arrows inferred the semiconductor VB top measured energies relative the common level of the oxide CB bottom.

4. Conclusion

In summary, the IPE/PC measurements have been used to determine the bandgap of ALD-grown $TaSiO_x$ layers consistently yielding the value 4.5±0.1 eV on three different substrates (Si, InP and $In_{0.53}Ga_{0.47}As$). Though the $TaSiO_x$ remains a good insulator thanks to high barriers provided by a SiO_x sub-network, on the In-containing substrates an In-rich IL is formed. The latter effectively reduces the interface barriers for electrons facilitating their injection and trapping in the silicate, recently revealed as the important source of charge instability of the $TaSiO_x$ -based insulating stacks [10].

References

[1] Radosavljevic M *et al* 2009 IEEE IEDM Tech. Dig. p 292
[2] Adelmann C *et al* 2011 Microelectron. Eng. **88** 1098
[3] Adamchuk V K *et al* 1992 Prog. Surf. Sci. **41** 111
[4] DiStefano T H *et al* 1974 J. Vac. Sci. Technol. **11** 1020
[5] Afanas'ev V V *et al* 2007 J. Appl. Phys. **102** 081301
[6] Chelikowsky J R *et al* 1976 Phys. Rev. B **14** 556
[7] Chou H Y *et al* 2010 Appl. Phys. Lett. **97** 132112
[8] Afanas'ev V V *et al* 2009 Microelectron. Eng. **86** 1550
[9] Walsh A *et al* 2008 Phys. Rev. Lett. **100** 167402
[10] Afanas'ev V V *et al* 2012 Appl. Phys. Lett. **100** 202104

IOP Conf. Series: Materials Science and Engineering **41** (2012) 012020 doi:10.1088/1757-899X/41/1/012020

Investigation of SiGe/Si heterostructures using state-of-the-art Auger Nanoprobes

P. K. Yadav [1, 2, *], E. Martinez[1], F. Bertin[1], M. Bouttemy[2], E. De Vito[3], O. Renault[1], J. M. Hartmann[1], A. Etcheberry[2] and A. Chabli[1]

[1]CEA, LETI, MINATEC Campus, 17 rue des Martyrs, 38054 Grenoble Cedex 9, France.
[2]Institut Lavoisier de Versailles, 45 av. des Etats-Unis, 78035 Versailles Cedex, France.
[3]CEA, LITEN, 17 rue des Martyrs, 38054 Grenoble Cedex 9, France.
*E-mail: *pradeep-kumar.yadav@cea.fr*

Abstract. The capabilities of nano-Auger were assessed for the characterization of SiGe multilayers epitaxially grown on Si(001) wafers. Reference sample consisting in stack of (300-500)-nm thick SiGe layers with a Ge content increasing in discrete steps from 6 to 30 % (as determined by X-ray diffraction) were used to that end. Composition measurements were performed on cross-sections with localized Auger Electron Spectroscopy using point analysis. The promising results obtained should enable in the near future high performance Auger mapping of real devices. The effect of native oxide removal either by argon sputtering or by HF etching was also addressed. Complementary results were otherwise obtained with Auger depth profiling using argon sputtering with Zalar rotation.

1. Introduction

SiGe/Si heterostructures are intensively used in order to boost the performances of ultra-short gate length complementary metal-oxide semiconductor (CMOS) devices [1-4]. For instance, intrinsic SiGe channels can be used to tune the threshold voltage and significantly increase the hole mobility in p-type Metal Oxide Semiconductor Field Effect Transistors.

The new generation state-of-the-art Auger nanoprobes with high lateral (<10 nm) and energy (0.1 %) resolutions make realistic the analysis of elemental and chemical composition at the deca-nanometer scale [5-6]. Indeed, the recent improvements in terms of spatial resolution (<8 nm) make Auger Electron Spectroscopy (AES) and Scanning Auger Microscopy (SAM) well suited for the localized chemical characterization of SiGe/Si heterostructures.

In this paper, the quantitative Auger point analysis and depth profiling are performed on stacks of (300-500)-nm thick SiGe layers with increasing Ge content taking advantage of the high energy resolution as well of lateral and depth resolutions of a state-of-the-art Auger nanoprobe.

2. Experimental Section

The Auger experiments have been performed using a PHI-700 Xi Auger nanoprobe system equipped with a co-axial cylindrical mirror analyzer (CMA) with energy resolution varying from 0.1 to 0.5 %. The measurements were carried out by using an electron beam voltage and current of 20 kV and 5 nA [~(8-10) nm spot size], respectively. Prior to Auger analysis, argon sputtering with Zalar rotation was carried out at 500 eV beam energy for 5 min to 20 min with an ion current of 280 nA to remove both the surface hydrocarbon contamination and the native oxide. Special care was taken to reduce sputtering-induced artifacts (such as ion milling and roughness enhancement) by using low incident energy. An alternative surface preparation was also tested based on the use of diluted (1 %) hydrofluoric acid to remove the native oxide. A short *in-situ* sputtering (750 eV for 30 s) was performed after the HF dip to remove the carbon contamination due to air exposure before loading in the Auger chamber. All the samples were examined under ultra-high vacuum conditions (6.5×10^{-9} mbar).

Epitaxial heterostructures consisting in stacks of (300-500)-nm thick SiGe layers with increasing Ge concentration from 6 to 30 % have been used as a reference samples. An Epi Centura Reduced Pressure – Chemical Vapour Deposition industrial cluster tool from Applied Materials was used to grow at high temperatures the SiGe multilayers on a p-type boron doped [~(7-10) Ohm·cm] (001) oriented Si substrate. The fabrication process has been described elsewhere [7]. The Ge concentrations were quantified by X-ray diffraction (XRD) measurements with the protocol described in Ref. [8]. Figure 1(a) shows a diagram of the SiGe/Si heterostructures.

A specific cross-section preparation based on medium energy ion milling was implemented in order to reduce surface roughness and topographical artifacts. We used the SM-09010 JEOL CP polisher after cutting the SiGe/Si multilayer to a suitable size (1 cm×1 cm) for the CP polisher. The CP tool sputters the sample perpendicularly to the top surface with an argon ion beam (6 keV, 60 mA and raster size 1 mm×300 μm). Figures 1(b) and 1(c) shows the 2-dimentional structure and optical microscope image of a cross-sectioned SiGe/Si sample.

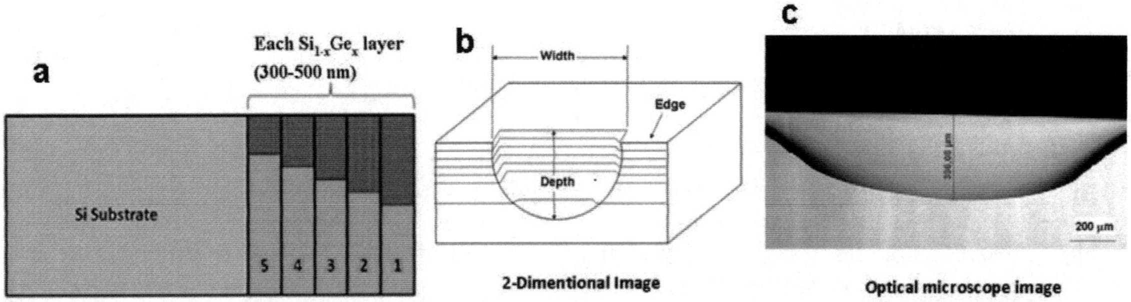

Figure 1. (a) Structure of a SiGe/Si multilayer. Ge concentrations measured by XRD on the $Si_{1-x}Ge_x$ sample where x =6.1% of Ge (point 5), x =12.3% (point 4), x =18% (point 3), x =23.5% (point 2) and x =28.8% (point 1). Localized cross-section preparation: **(b)** 2D schematic model of a CP cross section and **(c)** Optical microscope image of a cross sectioned SiGe/Si specimen.

3. Auger point analysis

Localized AES using point analysis which consists in acquiring a local Auger spectrum at a given point of the sample [~(8-10) nm spot size with 0.1 % energy resolution]. Here, it is performed on the sample cross- section to analyze the surface chemical composition after removal of both the native oxide and the hydrocarbon contamination. Figure 2(a) shows the Auger spectra obtained for each of the SiGe layers of the stack before any surface treatment. The carbon and oxygen (KLL) peaks present in all spectra are due to native oxide and surface contamination.

The Auger spectra obtained after argon sputtering for 20 min and after the surface chemical etching by 15-min dip in 1 % HF aqueous solution are shown in Figs. 2(b) and 2(c), respectively. Both surface preparations were successful for the lowest Ge concentrations. Indeed, the carbon or oxygen peaks almost disappear completely from the corresponding spectra. For the high Ge concentrations at the edge of the CP cross section, the carbon and oxygen contaminations are reduced but still present. This is probably due to a not perfectly flat surface at the edge. For the two surface treatments, the Si LVV, Ge LMM and

Si KLL peaks are detected throughout the stack. The germanium peak intensity decreases from point 1 to point 5 of the stack in agreement with the expected trend.

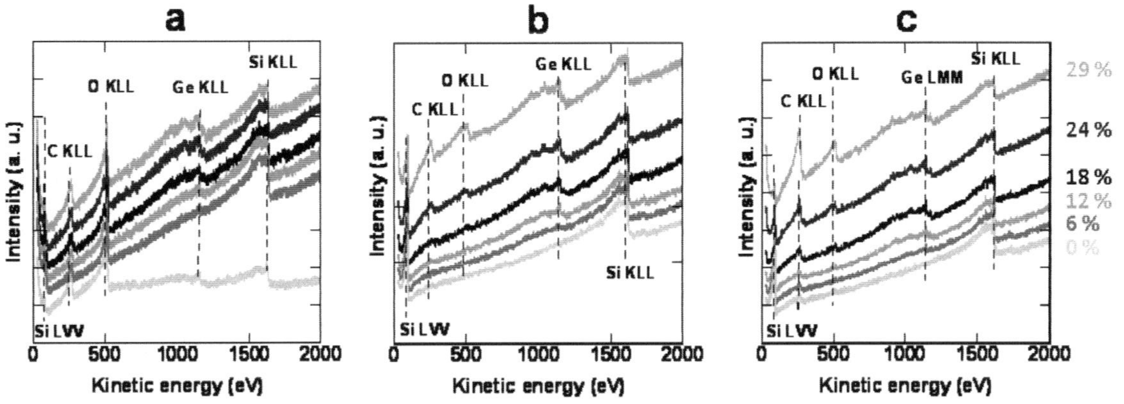

Figure 2. Auger point analysis on the cross-section of the SiGe/Si heterostructures (irradiation conditions: 20 kV, 5 nA and magnification: x 20 000). **(a)** Before surface treatment; **(b)** After argon sputtering (500 V, 280 nA for 20 minutes with Zalar rotation; and **(c)** After 1 % HF etching for 15 minutes followed by very short argon sputtering (500 V, 280 nA for 30 s).

For a quantitative analysis, the relative atomic concentration is estimated from the Si KLL (1614 eV) and Ge LMM (1143 eV) peaks intensities using the following equations:

$$C_{\mathrm{Si}} = \frac{I_{\mathrm{Si}} / S_{\mathrm{Si}}}{I_{\mathrm{Si}} / S_{\mathrm{Si}} + I_{\mathrm{Ge}} / S_{\mathrm{Ge}}}, \qquad C_{\mathrm{Ge}} = \frac{I_{\mathrm{Ge}} / S_{\mathrm{Ge}}}{I_{\mathrm{Si}} / S_{\mathrm{Si}} + I_{\mathrm{Ge}} / S_{\mathrm{Ge}}} \tag{1}$$

where C_{Si} and C_{Ge} are the atomic concentrations, I_{Si} and I_{Ge} are Auger peak intensities and S_{Si} and S_{Ge} are sensitivity factors, respectively for Si and Ge. The Auger peak intensities are extracted from the maximum amplitude of the Auger spectra derivative to remove the background signal. The resulting Si and Ge concentrations before and after surface preparation are shown in Figs. 3(a) and 3(b) together with XRD values. The results are given with a 15 % uncertainty using tabulated sensitivity factors (0.080 for Si KLL and 0.148 for Ge LMM [9]).

The significant differences between Auger and XRD concentrations prior to surface treatment may be explained by sensitivity factor errors. This is checked by using this time conventional Auger depth profiling. In this mode, the sample is sequentially sputtered by an Ar beam (5 keV, 5 μA) and analyzed by recording the Auger peak spectrum. The Si and Ge depth profiles are shown on Fig. 3(c). The presence of a carbon peak for small sputtering times confirms the presence of organics on the sample surface. The depth profiles are in good agreement with the expected values (from XRD). The use of the tabulated sensitivity factors is thus valid at 20 kV.

After argon sputtering, the Si and Ge concentrations are reasonably in agreement with those from XRD. However, a significant discrepancy is still observed for points 1 and 2. This is fully correlated with the limited efficiency of the Ar sputtering pointed out above close to the edge of the sample. In contrast, there is a very good agreement between XRD and Auger concentrations after the HF etching, whatever the layer of the stack. These results show that sample surface preparation using HF leads to reliable measurement by nano-Auger point analysis of Si or Ge relative concentrations atleast in the investigated concentration range.

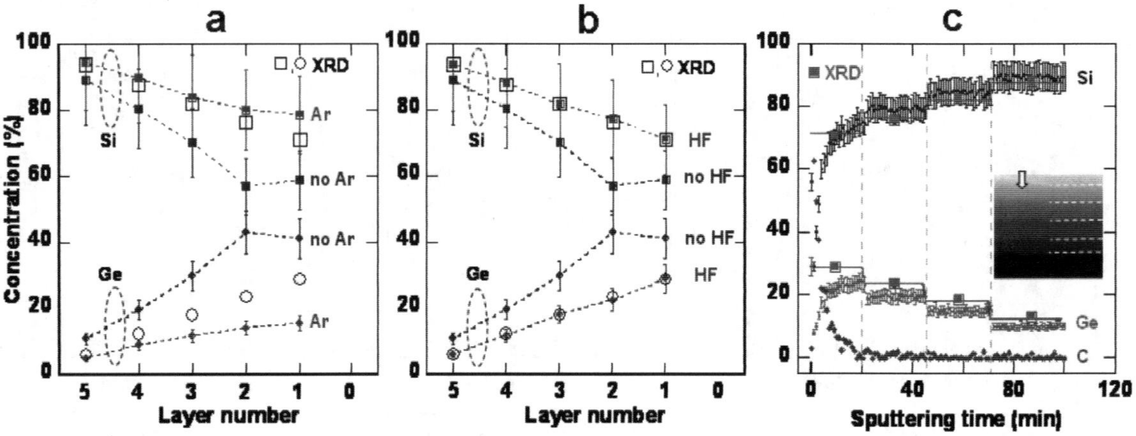

Figure 3. Comparison of Auger and XRD concentrations **(a)** before (**no Ar**) and after (**Ar**) Ar sputtering; **(b)** before (**no HF**) and after (**HF**) HF etching; **(c)** Auger Depth Profiling on the SiGe/Si heterostructure (irradiation conditions: 20 kV, 5 nA, magnification: ×20 000) and Ar Sputtering (5 kV and 5 μA) with Zalar rotation (inset: SEM stack view).

4. Conclusion

The capabilities of nanoAuger for the characterization of SiGe/Si has been investigated using reference samples consisting in stacks of SiGe layers epitaxially grown on Si wafers with a Ge concentration staircase increasing from 6 to 30 % (as determined by X-ray diffraction). Auger depth profiling validates the use of the tabulated sensitivity factors up to 20 kV. Tests of SiGe composition measurements were performed on cross-sections using Auger point analysis with different surface preparation. The limited efficiency of the Ar sputtering results in a significant discrepancy between Auger and XRD concentrations. In contrast, the sample surface preparation using 1% Hydrofluoric acid leads to reliable measurement of Si or Ge relative concentrations in the investigated concentration range.

Acknowledgments
This work was supported by the "European Euro Talent" project and by the French National Research Agency (ANR) through the "Recherche Technologique de Base" Program.

References
[1] Mizuno T. et al., 2003 *IEEE Trans. Electron Devices* **50** 988.
[2] Packan P. et al., Proceedings of the 2009 IEDM Conference, p. 659.
[3] Le Royer C. et al., 2011 *Solid State Electronics* **65-66** 9.
[4] Le Royer C. et al., Proceedings of the 2011 IEDM Conference, p. 391.
[5] Karpuzov D. et al., 2006 *Surf. Interface Anal.* **38** 287.
[6] Yadav P. et al., 2011 AIP Conf. Proc. **1395** 113.
[7] Hartmann J M. et al., 2006 *ECS Transactions* **3** 219.
[8] Destefanis V. et al., 2009 J. Cryst. Growth **311** 1070.
[9] Hedberg C L, *Handbook of Auger Electron Spectroscopy, 3rd Edition* **1995**: Physical Electronics, Inc.

IOP Conf. Series: Materials Science and Engineering **41** (2012) 012021 doi:10.1088/1757-899X/41/1/012021

Electron spin resonance study of point defects in thermal GaAs/GaAs-oxide structures

S. Nguyen[1], V. V. Afanas'ev[1], and A. Stesmans[1,2]

[1] Semiconductor Physics Laboratory, Department of Physics, University of Leuven, 3001 Leuven, Belgium

[2] E-mail: andre.stesmans@fys.kuleuven.be

Abstract. In an attempt to atomically assess interface and oxide-related point defects, a first basic multifrequency low-temperature electron spin resonance study has been carried out on semi-insulating (Fe compensated) GaAs/oxide structures, implying both powders and (100)GaAs/oxide slices, thermally grown in the range T_{ox}=350-615 °C. Various spectra are observed: As for powders, this includes the 4-line EL2 defect spectrum centered at g~2.043 and characterized by the isotropic hyperfine constant A_{iso}~ 910 G, ascribed to the $^{75}As_{Ga}^{+}$ antisite defect. Observed in freshly crushed powder, it substantially increases in density with oxidation, in line with theoretical expectation; A maximum appears reached for at T_{ox}~440 °C. It is not observed in the parent c-GaAs wafer. A second isotropic signal is observed at g~1.937 in powders for T_{ox} in the range 510-615 °C, but only after additional VUV irradiation; it may correspond to As clusters. In bulk (100)GaAs, we observe the 5-branch spectrum of substitutional Fe impurities (spin S=5/2) in GaAs, with inferred crystal field constant a≈360 G, well in line with previous observations. The results are discussed within the framework of advanced theoretical interface and defect models and previous experimental assessment.

1. Introduction

The introduction of III-V semiconductors, such as GaAs, as high-mobility channels in field effect transistors has been hindered by the presence of large density of electron states (defects, traps) at the GaAs/oxide interface [1] to such extend that it leads to strong Fermi-level pinning at the oxide/III-V semiconductor interface, the origin of which is not fully understood nor atomically identified and quantified. Though investigated over many decades, this lack of adequate interface (surface) passivation has been a main obstacle on the route to development of high-performance III-V MOS technology. One solution offered has been the use of $(Ga,Gd)_2O_3$ dielectrics grown by means of ultra-high vacuum molecular beam epitaxy as gate oxide [2]; Fermi level unpinning has also been realized by applying an Al_2O_3 dielectric layer directly on (100)GaAs by means of atomic layer deposition.[3, 4] The unprecedented efforts devoted to the search for the functional replacement of the conventional SiO_2 insulator in Si-based CMOS IC technology by a dielectric of higher dielectric constant (κ) has resulted in an outstandingly broad knowledge, experience, and understanding of deposited higher-κ insulators for application in Si MOS devices. In its wake, this progress has refocused the interest in surmounting the key problem with the interface passivation of III-V semiconductors through the potential application of alternative (higher-κ) deposited gate dielectrics for which now a realm of choices has become available (e.g., Al_2O_3, HfO_2, $LaAlO_3$).

IOP Conf. Series: Materials Science and Engineering **41** (2012) 012021 doi:10.1088/1757-899X/41/1/012021

Still, a basic question concerns the atomic origin of the interface states that cause the Fermi level pinning. In citing a conclusion from recent theoretical work, "these states must be native defect states of the semiconductor", just as is the case for Si [5]. Many suggestions have been made, largely of unknown or hypothetical atomic nature.

On theoretical grounds, various types of native point defects such as interfacial As-As dimers (like-atom bonds), AsGa antisites, Ga dangling bonds (DBs), and elemental As, have been projected as potential origin of detrimental traps with levels in the GaAs band gap. Recent theory has concluded that problematic III-V gap states are caused by interfacial As-As bonds [5,6], which presence has been demonstrated experimentally.[7] Pertinently, according to Gibbs free energy considerations, Ga has a much stronger affinity for oxygen than As, so during thermal oxidation or post oxidation treatment at temperatures ($\gtrsim$ 500 °C) As-oxides become less abundant leaving a local excess in As.[1, 4, 8]

Little is known regarding the precise atomic nature on verified experimental grounds. Investigation of the chemical bonding at surfaces of InxGa1-xAs systems (for x ranging from 0 $\rightarrow$ 1) with native oxides and atomic layer deposited Al_2O_3 oxides by means of X-ray photoelectron spectroscopy (XPS) has concluded that elimination of interfacial As-oxides is a key requisite for Ef unpinning [4,7]. The work also reports the presence of As^0 (As-As bonding) and Ga^{3+} states, that would originate interface defects states near the semiconductor midgap and conduction band (CB), respectively.

Thus, as to the injurious interface states, atomistic identification of the originating defects remains a challenge. As has been illustrated before for the interfaces of SiO_2 with Si [9] and Ge_xSi_{1-x} [10], ESR has emerged as the tool of choice for such task, essentially enabled through the method's inherent potential of mapping a point defect's unpaired electron wave function -this unlike other optical (absorption) and electrical (e.g., deep level transient spectroscopy) techniques

The literature on the study of defect by conventional ESR studies on GaAs, or III-V compounds in general, is rather limited. As to GaAs, these are limited to "bulk" (including deposited layers) properties. One reason here is the difficulty with detection as the ESR signals are generally strongly broadened due to the presence of unresolved hyperfine interactions and subject to strong multiple hyperfine splitting. Without doubt, the most studied defect in GaAs is the native As_{Ga}^+ antisite defect (EL2 center) in which an arsenic ion substitutes for a gallium ion in GaAs. As detected by ESR, the center shows an isotropic 4-peak spectrum characterized by g≈2.037 and isotropic hyperfine interaction constant $A_{iso} \approx 947$ G. [11-15]; the model is supported by theoretical insight [16]. Recently, there have been theoretical efforts focused on the As_{Ga} as an interface defect [17-19]. These works conclude that the oxidation treatment can induce (near) interface As_{Ga} antisite defects as well as As and Ga dangling bonds [6]. Scarrozza et al. [18] have built a model of As antisite creation during the oxidation process of GaAs. Because the much stronger affinity of Ga than As for oxygen, the oxidation may lead to As_{Ga} antisite creation at the interface.

Lin and Robertson [6] analyzed As dimer levels in semiconductor bands. The bonding (σ) state of the As dimer is in the valence band, so the ESR method can detect only the antibonding (σ*) state lying right under the conduction band.

Taking a basic approach, in this work we report on a first multi-frequency low temperature ESR study of GaAs/GaAs-oxide entities, with main focus on tracing paramagnetic interface point defects (traps) and, in this way, to come to atomic identification of electrically relevant traps thus aiming for experimental verification of theoretical hypotheses. So far, no ESR-active centers have been reported. To boost ESR sensitivity, in an initial step most experiments are carried out on thermally oxidized c-GaAs powders -this to enhance the effective interface area per sample volume. At the same time, this enhances the volume ratio of GaAs oxide-to-GaAs material much relevant for tracing (near-interface) point defects in the oxide. As a first major finding, thermal oxidation is found to result in a substantial increase in As_{Ga}^+ antisite defects.

To assess the atomic nature of the interface on atomic scale in terms of occurring point defects.

IOP Conf. Series: Materials Science and Engineering **41** (2012) 012021 doi:10.1088/1757-899X/41/1/012021

2. Experimental details

A. Samples

Samples studied were manufactured starting from commercial single side polished semi-insulating (Fe-compensated) 2-inch diam. (100)GaAs wafers about 470 μm thick. From these, generally, two groups of samples were prepared for ESR analysis. For the major part, to maximize the effective semiconductor/oxide interface area in a (size-restricted) ESR sample, GaAs powders were made through pulverizing in room ambient pieces of GaAs wafer by means of the mortar-and-pestle (both of pure Al_2O_3) method, initially "cleaned" through crunching of sacrificial pieces of GaAs. This resulted in powders of mean particle size $d_m \approx 5$ μm, with typical distribution standard deviation $\sigma \approx 3$ μm. Subsequently, sets of powder were thermally oxidized (O_2: 99.9999 %; 1.1 atm) at various temperatures in the range T_{an} = 350 - 615 °C for various times (22 - 390 min). The oxidation procedure was halted by rapid cooling to room temperature (RT) in unaltered (O_2) sample ambient by offsetting the conventional electronically stabilized resistance-heated furnace from the double-walled quartz insert, housing the samples to attained oxide thickness d_{ox}, oxidation at 500 °C for 22 min (parabolic growth range) typically results in $d_{ox} \approx 15$ nm [20].

After a first ESR test, some samples were additionally subjected to irradiation for prolonged times (30 - 120 min), with frequent intermittent stirring of the exposed powder, by unbiased vacuum UV (VUV) 10.02 eV photons (flux ~10^{15} $cm^{-2}s^{-1}$) -this with the intend to maximally enhance/activate the density of paramagnetic (ESR-active) point defects. Such step may unveil strained and/or weak bonds (bond rupture) and diamagnetic precursor sites [21].

In a second, more conventional approach for studying the (potential) magnetic field anisotropy of ESR-active interfacial point defects, oxidations were carried out in the range 430 – 555 °C on various sets of slices (typically 10-20) of 2 x 10 mm^2 main area cut from the main wafers, with their long edge along a <$1\bar{1}0$> direction. A typical ESR sample was comprised of a stack of 10-15 slices.

B. Electron spin resonance spectroscopy

Conventional absorption mode ESR observations were carried out at 4.2 K using X (~8.9 GHz), K (~20.5 GHz), or Q-band (~33.5 GHz) spectrometers driven in the adiabatic slow passage mode, as described elsewhere [22], with the applied magnetic field **B**, at angle φ_B with the [100] sample normal **n**, rotating in the ($1\bar{1}0$) plane. Usually, first derivative absorption (dP_μ/dB) signals were detected by employing sinusoidal modulation (~100 kHz) of **B**. The modulation amplitude B_m and the incident microwave power P_μ were appropriately reduced to avoid any visible signal distortion. Spectral averaging was applied (up to ~100 scans) to enhance signals. Additionally, Q-band saturation spectroscopy –high-power dispersion-like (rapid passage) first harmonic and second harmonic phase-quadrature saturation mode [see, e.g., Ref. 23] were applied to enhance detectivity for some signals and to assist in disentanglement of overlapping signals and probe signal shapes. As empirically detailed elsewhere, it appears that when appropriate spectrometer settings are met for a certain spin center studied, the second harmonic phase-quadrature (out-of-phase) spectra emerge as identical to the defect's absorption spectrum [23, 24].

Defect densities were determined through double numerical integration of the as-observed as well as computer-simulated dP_μ/dB signals, by making use of a calibrated co-mounted Si:P marker [g(4.2 K) = 1.99869, spin S = ½], also serving as g marker.

3. Experimental results and interpretation

As mentioned, in the search for signals, the initial part of the ESR study was carried out on c-GaAs powders, the results of which will be the subject of this report. Thus far, most ESR data for this work were taken using the Q-band spectrometer because, as compared to the other spectrometers operated, the former one allows for the most versatile accomplishment of potentially ESR-sensitivity enhancing detection modes, i.e., 1st harmonic dispersion and 2nd harmonic phase-quadrature saturation spectroscopy -a key element in the search for unknown ESR signals.

A. As_{Ga} antisite defect

Obviously, in crunching high-purity c-GaAs, there is the risk of inadvertent introduction of impurities. Indeed, that extreme care is mandatory is illustrated by Fig. 1, showing a regular absorption mode dP_μ/dB K-band spectrum, centered around the g = 2 spot, measured on the firstly prepared as-crushed GaAs powder. Besides the Si:P marker signal, a prominent signal is observed at $g_c = 2.0034 \pm 0.0001$, characterized by a peak-to-peak width $\Delta B_{pp} = 7.7$ G, a Lorentzian lineshape, and low saturability - without doubt stemming from C impurification [25]. For this "number-one" (first made) GaAs powder, the C contamination has likely resulted from remnants of the final rinsing (acetone) step in the initial cleaning procedure of the mortar-and-pestle tool. After some initial sacrificial GaAs pulverization preparations ("tool training"), the signal remains absent for all further powders made - the ones studied henceforth in this work.

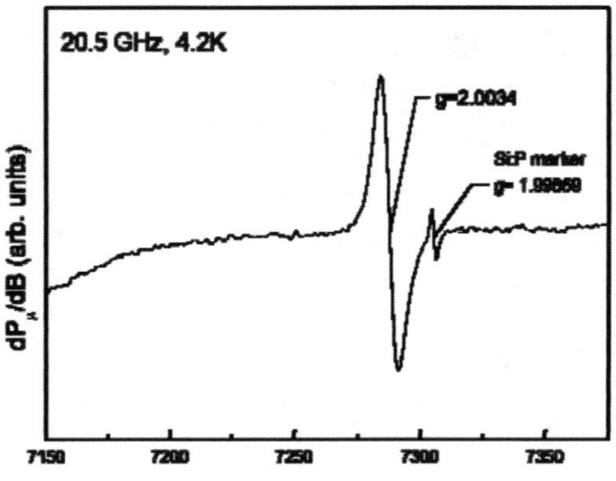

Figure 1: Conventional first-derivative absorption K-band spectrum observed, using $P_\mu = 2.5$ μW; $B_m = 1$ G, on the first manufactured GaAs powder sample showing the observation of a prominent signal at g = 2.0034 stemming from C contamination. The signal at g = 1.99869 originates from a co-mounted marker sample.

Figure 2 shows a comparison of broad-field scan Q-band dispersion-mode like spectra observed on as-crushed and thermally oxidized GaAs powders, both before and after an additional VUV irradiation step. Various signals are observed, among which some of spectrometer background origin, and a signal related to the Fe dopant in the parent GaAs wafer (*vide infra*). Starting with the as-crushed state, though partially overlapped by the Fe signal, perhaps most eye catching is the observation in the 1 - 1.4 T range around the g = 2.0 field position, of a 4-peak pattern: Three peaks are clearly resolved, the fourth one being overwhelmed by other signals, with a splitting of ~ 2740 G between the outer peaks. Further perusal of the spectrum, in comparison with literature data, readily revealed it to concern the As_{Ga}^+ (EL2) antisite spectrum, well known from previous ESR studies [11-15] on various types of "bulk" GaAs. The experimental spectrum may well be described by the isotropic spin Hamiltonian for an electron spin S=1/2

$$\mathcal{H} = g\mu_B\mathbf{B.S} + A_{iso}\mathbf{I.S}, \qquad (1)$$

comprised of the electron Zeeman and hyperfine (hf) interaction terms, respectively, where the latter describes hf interaction with a single nucleus (site) of nuclear spin I = 3/2 and 100 % natural abundance (i.e., ^{75}As). Spectra simulation gives for the Hamiltonian parameters g = 2.043 and isotropic hf interaction constant $A_{iso} = 910 \pm 10$ G, well in agreement with previous results on "bulk" c-GaAs [14,15]. Importantly, despite intensive efforts, no such signal could be observed on a stack of

(100)GaAs substrate slices; the signal observed in the as-crushed powder is induced by the pulverisation treatment.

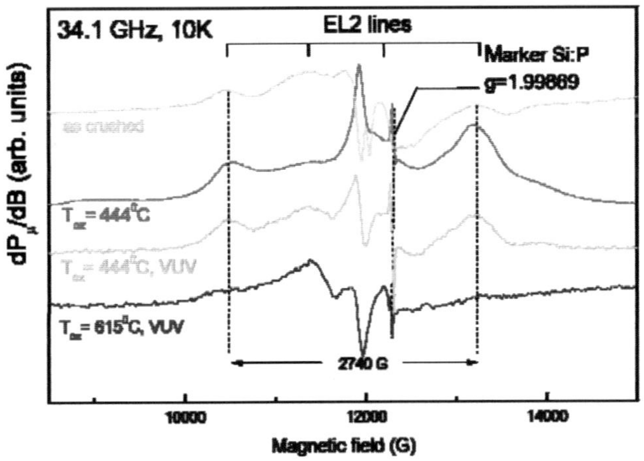

Figure 2: High-power (P_μ = 1.7 mW) dispersion-like mode spectra observed on as-crushed GaAs powder and powders thermally oxidized at two different temperatures, as indicated, in combination with additional post-oxidation VUV (10.02 eV) irradiation (~1 h). This shows the observation, as undifferentiated absorption-like peaks, of the typical EL2 quartet spectrum (as marked) centered at g = 2.043, which is ascribed to the As_{Ga}^{+} antisite defect.

Next, turning to the oxidation step, a most notable observation is that thermal oxidation at 444 °C results in a substantial enhancement of the EL2 signal. Yet, additional VUV irradiation does not seem to have much influence on the signal, perhaps causing some weak decrease. However, the signal disappears again, or at least is drastically reduced, when oxidizing at 615 °C. Though not all data are collected in Fig. 2, the overall evolution of the EL2 appearance in the covered T_{ox} = 350 °C - 615 °C range saturation is clear: Upon thermal oxidation of as-crushed GaAs powder from T_{ox} = 350 °C onward, the EL2 signal is enhanced, reaches a maximum with increasing T_{ox}, to be drastically reduced again for T_{ox} approaching 615 °C. At this point, providing information on the absolute density of occurring EL2 centra appears less straightforward: Though well useful for studies on relative basis, the currently applied dispersion-like defect probing method is not quite fit for reliable determination of absolute defect densities. For this, instead, regular unsaturated absorption ESR spectra would be needed, which measurement suffer from inferior sensitivity.

The above revealed striking evolution in EL2 center appearance merits further attention. First, we thus find that the As_{Ga}^{+} antisite defect, not observable in the parent c-GaAs wafer, is generated to some extend by GaAs crushing. In a first straightforward interpretation, this may be seen as simply due to mechanically induced damage to the c-GaAs material imposed during crushing. However, as hinted by the subsequent oxidation data, it cannot be excluded it to have resulted (to some extent) from some native oxide formation in room ambient during the pulverization procedure. Next, as a prime observation, the observed substantial increase in EL2, a c-GaAs bound defect, in growing density with T_{ox} increasing in the range 350 - 450 °C points to the enhanced As_{Ga}^{+} antisite formation at the GaAs/oxide interface, or put in other nuance, it represents atomistic identification of a particular kind of interfacial like-atom As-As bond formation. The result is in line with previous finding [7] and theoretical insight. [6,17] Within the acquired understanding of GaAs oxidaton, it may be seen as the result of the fact that during thermal oxidation, the much stronger affinity of Ga than As for oxygen leads to predominant Ga-oxide formation, leading to As enrichment at the interface. As a solid result, the data thus clearly reveal interfacial As enrichment as epidemic for thermal oxidation [5, 7, 19] in the range $T_{ox} \lesssim 500$ °C. For higher oxidation temperatures then, where the EL2 density is found to decline again, the reason may be that the diffusivity of As through the formed GaAs oxide may increase to such extend as to enable continuously balanced (exhaustion) As escape through evaporation [26]. As reported before, for thermal treatments at T $\gtrsim$ 500 °C, the oxide formed is almost completely Ga_2O_3 [8, 27]. Along these lines, the fading of the EL2 signal for $T_{ox} \rightarrow 615$ °C is almost certainly related to the predominant formation of the polycrystalline Ga_2O_3 phase, likely the chemically and thermally stable β-Ga_2O_3 phase. However, in fairness, in a totally different view it cannot be excluded that the increase in T_{ox} results in larger As-cluster formation at the interface, apparently so far inaccessible by ESR.

B. g = 1.937 signal

Turning to observations made by Q-band 2nd harmonic saturation spectroscopy, Fig. 3 shows some spectra measured at 10 K on GaAs powder samples oxidized at 510 °C or 615 °C. For T_{ox} = 510 °C, we notice that additional VUV irradiation results in ESR activation of a prominent signal characterized by g = 1.937 ± 0.002 and width (at half height) ΔB = 158 ± 3 G; it not observed in the as-oxidized state. As show by the bottom spectrum (equal VUV treatment), the signal is much reduced for T_{ox} increasing to 615 °C. In terms of applied oxidation temperatures, the coming and going of the signal much resembles that of the EL2 signal behavior, perhaps suggesting some correlation in generation modalities. In support, we also add that the signal could only be observed in oxidized powdered GaAs, not in oxidized c-GaAs slices.

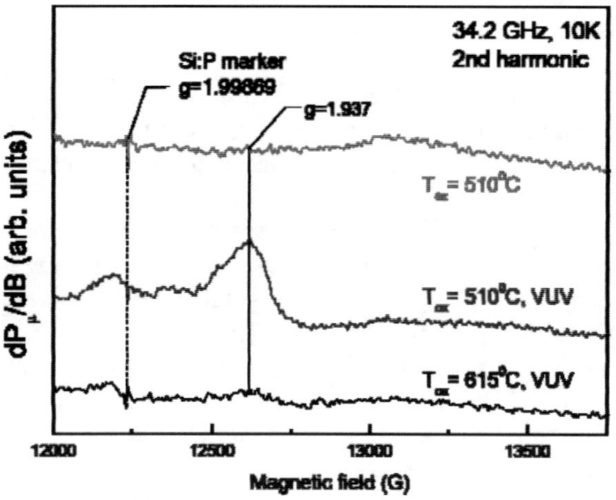

Figure 3: Q-band 2nd harmonic phase-quadrature saturation spectra observed over the g = 2.036 - 1.777 range, using P_μ = 1.7 mW and B_m = 1.5 G, on c-GaAs powder oxidized at 510 or 615 °C. The spectra illustrate the appearance of a signal at g = 1.937 under additional VUV (10.02 eV) irradiation.

The signal is also observed by conventional K-band ESR, but is now found to be drastically narrowed to $\Delta B_{pp} \approx 52$ G, which points to a dominant inhomogeneous line broadening mechanism, perhaps g-spread broadening. As for the atomic origin of this signal, in the absence of the observation of any correlated resolved hf structure, suggestions can only be highly hypothetical. Pertinently, in scanning the literature, we note that the observation of an ESR signal with similar g value (g = 1.94) has been reported for semi-insulating as-grown GaAs after prolonged irradiation by 1.18 eV photons [28]; However, no atomic model was suggested. Perhaps, the As clusters may be advanced as possible defect candidate.

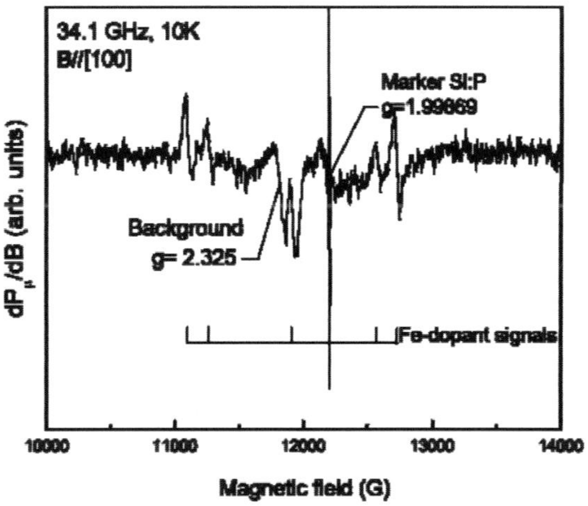

Figure 4: Conventional Q-band derivative absorption spectrum observed at 10 K using $P_{\mu} = 17$ μW on a stack of (100)GaAs slices for **B**//**n** ([100] sample normal) showing the quintet spectrum, centered at g = 2.043, originating from substitutional Fe dopants in the GaAs crystal. The additional signal at g = 2.325 originates from cavity background.

Finally, over the many ESR measurements made in the current work, we should add that more ESR signals have been traced, still in need of more in depth study. One of these is the observation of the substitutional Fe compensation dopant in the parent GaAs substrate, of which a conventional Q-band absorption spectrum observed for **B**//[100] on a pristine (100)GaAs slice is shown in Fig. 4; It is composed of a quintet, centered at g ~ 2.07, the anisotropic splitting resulting from crystal field interaction. The inferred crystal field constant $a \approx 360$ G is well in line with previous observations [29]. More details will be presented elsewhere.

4. Conclusion

Results have been presented of a first extensive low-temperature multifrequency ESR study of thermally oxidized GaAs. For ESR sensitivity reasons, this is in first instance mainly carried out on mechanically pulverized (powder) of device grade c-GaAs material, with the intend to assess occurring paramagnetic point defects potentially operating as (interfacial) trapping/recombination centers, and in this way unveiling their true atomic nature through the ESR-specific defect identification potential. Several signals are observed, including the As_{Ga}^{+} (EL2) antisite defect, the compensation Fe dopant in the GaAs substrate material, and a center at $g_c = 1.94$; The latter is maximally observed in GaAs powders oxidized at $T_{ox} \sim 510$ °C, but only after additional VUV irradiation. As a first hypothesis, it may concern As clusters.

A most noteworthy result is that As_{Ga}^{+} centers are found to be substantially generated by thermal oxidation in increasing densities with increasing T_{ox} from ~310 °C onward, herewith revealing interfacial As enrichment, in agreement with theoretical insight. For higher T_{ox} ($\gtrsim$ 615 °C), its generation drops off. As the EL2 center is known to be a deep (double) level center [30], its generation provides one account for the (known) inferior interface quality attained through simple thermal oxidation of GaAs; Thermal oxidation, at least in the range $T_{ox} \lesssim 550$ °C, is to be avoided.

5. Acknowledgments

The authors are indebted to Dr. G. Brammertz for the supply of the basic GaAs material.

6. References

[1] C. W. Wilmsen, in *Physics and Chemistry of III-V Compound Semiconductor Interfaces*, edited by C.W. Wilmsen (Plenum, New York, 1985).
[2] M. Passlack, M. Hong, J. P. Mannaerts, R. L. Opila, S. N. G. Chu, N. Moriya, F. Ren, and J. R. Kwo, IEEE Trans. Electron Devices **44**, 214 (1997)
[3] P. D. Ye, G. D. Wilk, B. Yang, J. Kwo, S. N. G. Chu, S. Nakahara, H.-J.L Gossmann, J. P. Mannaerts, M. Hong, K. K. Ng, and J. Bude, Appl. Phys. Lett. **83**, 180 (2003)
[4] C. L. Hinkle, E. M. Vogel, R. M. Wallace, and P. D. Ye., Current Opinion Solid State Mat. Sci. **501** (2011)
[5] J. Robertson, Microelectron. Eng. **86**, 1558 (2009)
[6] J. Robertson and L. Lin, Microelectron. Eng. **88**, 1440 (2011); L. Lin and J. Robertson, Appl. Phys. Lett. **98**, 082903 (2011).
[7] W. Wang, C. L. Hinkle, E. M. Vogel, K. Cho, and R. M. Wallace, Microelectron. Eng. **88**, 1061 (2011).
[8] O. R. Monteiro and J. W. Evans, J. Vac. Sci. Technol. A7, **49** (1989)
[9] E. H. Poindexter, P. Caplan, B. Deal, and R. R. Razouk, J. Appl; Phys. **50**, 5847 (1979); K. L. Brower, Appl. Phys. Lett. 43, 1111 (1983)
[10] A. Stesmans, P. Somers, and V; V. Afanas'ev, Phys. Rev. B**79**, 195301 (2009)
[11] R. J. Wagner, J. J. Krebs, G. H. Strauss, and A. M. White, Solid State Commun. **36**, 15 (1980)
[12] A. Deville, B. Gaillard, K. Sedeek, H. Carchano and K. W. H. Steven, J. Phys. Condens. Matter **1**, 9369 (1989).
[13] H. J. von Bardeleben, D. Stievenard, D. Deresmes, A. Huber and J. C. Bourgoin, Phys. Rev. B **34**, 10 (1986).
[14] H.-J Sun, G. D. Watkins, F. C. Rong, L. Fotiadis, and E. H. Poindexter, Appl. Phys. Lett. **60**, 718 (1992)
[15] W. Jost, M. Kunzer, U. Kaufmann, K. Köhler, J. Schneider, and H. C. Alt, Semicond. Sci. Technol. **7**, 1386 (1992)
[16] M. C. Esteves, A. B. Rocha, N. V. Vugman, and C. E. Bielschowsky, Chem. Phys. Lett. 453, 188 (2008)
[17] H. Overhof and J. M. Spaeth, Phys. Rev. B **72**, 115205 (2005).
[18] M. Scarrozza, G. Pourtois, M. Houssa, M. Caymax, A. Stesmans, M. Meuris, and M. M. Heyns, Appl. Phys. Lett. **95**, 253504 (2009)
[19] J. Robertson, Appl. Phys. Lett. **94**, 152104 (2009).
[20] D. N. Butcher and B. J. Sealy, J. Phys. D: Appl. Phys. **11**, 1451 (1978)
[21] A. Stesmans and V. V. Afanas'ev, Appl. Phys. Lett. **69**, 2056 (1996)
[22] A. Stesmans, B. Nouwen, and K. Iakoubovskii, J. Phys.: Condens. Matter **12**, 7807 (2000)
[23] D. L. Griscom, Phys. Rev. B **20**, 1823 (1979): T. E. Tsai and D. L. Griscom, J. Non-Cryst. Solids **91**, 170 (1987)
[24] R. Fuller, H. Evans, C. Gamten, B. Czagas, M. Morrison, D. Decrosta, R. Lowry, P. Lenahan, and C. Frye, IEEE Trans. Nucl. Sci. **43**, 2565 (1996)
[25] J. G. Gambino and J. A. Thompson, Solid State Commun. **34**, 15 (1980); A.I. Shames, A.M. Panich, E. Mogilko, J. Grinblat, E.W. Prilutskiy, and E.A. Katz , Diamond Relat. Mater. **16**, 2039 (2007)
[26] G. P. Schwartz, G. J. Gualtieri, G. W. Kammlott, and B. Schwartz, J. Electrochem. Soc. **126**, 1737 (1979)
[27] R. Korbutowicz and J. Prazmowska, in *Semiconductor Technologies*, edited by J. Grym (InTech, 2010). p. 105.
[28] M. Baeumler, U. Kaufmann, and J. Windscheif, Mat. Res. Soc. Symp. Proc. Vol. **48**, 201 (1985)
[29] M. de Wit and T. L. Estle, Phys. Rev. **132**, 1 (1963)
[30] J. Dabrowski and M. Scheffler, Phys. Rev. B **40**, 10 391 (1989)

Biostability of an implantable glucose sensor chip

M Fröhlich, M Birkholz[1], KE Ehwald, P Kulse, O Fursenko, J Katzer
IHP, Im Technologiepark 25, 15236 Frankfurt (Oder), Germany

E-mail: birkholz@ihp-microelectronics.com

Abstract. Surface materials of an implantable microelectronic chip intended for medical applications were evaluated with respect to their long-term stability in bio-environments. The sensor chip shall apply in a glucose monitor by operating as a microviscosimeter according to the principle of affinity viscosimetry. A monolithic integration of a microelectromechanical system (MEMS) into the sensor chip was successfully performed in a combined 0.25 µm CMOS/BiCMOS technology. In order to study material durability and biostability of the surfaces, sensor chips were exposed to various *in vitro* and *in vivo* tests. Corrosional damage of SiON, SiO_2 and TiN surfaces was investigated by optical microscopy, ellipsometry and AFM. The results served for optimizing the Back-end-of-Line (BEoL) stack, from which the MEMS was prepared. Corrosion of metal lines could significantly be reduced by improving the top-most passivation layer. The experiments revealed no visible damage of the actuator or other functionally important MEMS elements. Sensor chips were also exposed to human body fluid for three month by implantation into the abdomen of a volunteer. Only small effects were observed for layer thickness and R_a roughness after explantation. In particular, TiN as used for the actuator beam showed no degradation by biocorrosion. The highest degradation rate of about 50 nm per month was revealed for the SiON passivation layer. These results suggest that the sensor chip may safely operate in subcutaneous tissue for a period of several months.

1. Introduction

The application of microelectronics in the field of biosensorics requires the consideration of certain constraints with respect to the usage of technical materials in corrosive environments [1, 2]. We report here on a sensor chip for the continuous monitoring of glucose operating by the principle of affinity viscosimetry [3, 4]. The development aims at an implantable device for diabetes diagnostics and therapy that shall stay in human subcutaneous tissue for a time span of months. The sensor operates by determining the viscosity of a sensoric liquid that varies in accordance with glucose concentration in the surrounding tissue [5, 6]. The sensor chip is realized as a microelectromechanical system (MEMS) operating with a mechanically bendable beam in an electrostatic actuation scheme and from the bending velocity of which the viscosity can be derived [7].

In contrast to the usual procedure with MEMS being fabricated from the Si bulk, the microvis-cosimeter presented in this work has been prepared from the Backend-of-line (BEoL) stack [8]. The functional layers staying in contact with the bio-milieu are thus the surface passivation from SiON (Pas), the interlayer dielectric (ILD) from SiO_2 as well as the bendable beam and the ground plate electrode made from TiN. Figure 1 displays a section from the unfilled MEMS cavity that will be

[1] To whom any correspondence should be addressed.

filled with sensoric liquid during operation. Parts of the beam and ground plate in the vicinity of an anchor point can be recognized that both are fabricated from metallically conductive TiN in order to enable the electrostatic actuation scheme of the BioMEMS. TiN has advantageously been chosen, since the material is well established in semiconductor fabrication technology, where it is used as an anti-diffusion layer below and above lateral interconnects from Cu-alloyed aluminium Al:Cu. The slope observable for the SiO_2 ILD layer has been introduced during etching of the cavity for suspending the TiN beam according to the chronological progression of the etch front, where top layers were subjected to the etchant for a longer time. The top-most layers is formed by the SiON passivation exhibiting a lower etch rate than SiO_2 and forming an overhang after the etch process that stopped on the TiN ground plate.

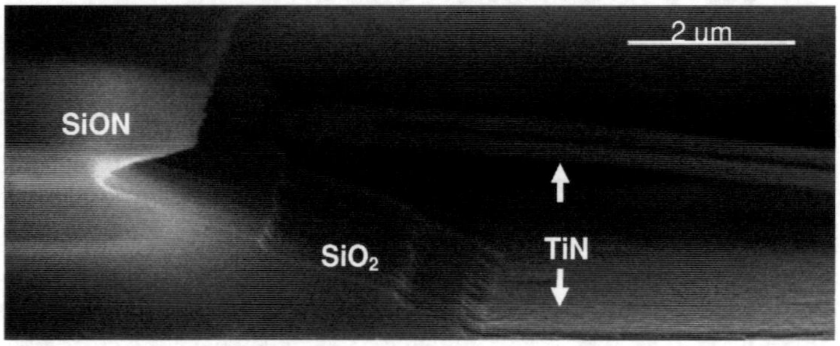

Figure 1. Scanning electron micrograph of a part of the Bio-MEMS displaying the different materials staying in contact with body fluid after implantation of the glucose sensor chip.

Previous investigations revealed TiN to be sufficiently biostable for the applications intended here [9, 10]. The focus of this work was thus on ILD and Pas layers, for which *in vitro* and *in vivo* loss rates should be determined in order to derive a first estimate for a possible life time of implanted biochips in the body tissue.

2. Materials and methods

2.1. Sample preparation

Sensor chips were prepared on 200 mm CZ-Si wafers in a 0.25 µm CMOS/BiCMOS technology [11]. Different variants of the so-called value technology are available with the base variant supplying metal layers M1, M2 and M3 from Al:Cu and sheet resistances of 84, 55 and 55 mΩ/sq embedded in four isolating SiO_2 layers ILD0..ILD3. Furthermore, top-metal layers TM1 and TM2 (16 and 10 mΩ/sq) and additional ILD4 and ILD5 layers may be supplemented, yielding an increased geometrical height of 8 µm of the full BEoL stack. Planarizations of intermediate BEoL stacks are performed in the standard flow after ILD1 to ILD3 by chemical-mechanical polishing (CMP), but not after ILD4 and ILD5.

MEMS actuators were prepared as fourfold clamped X-shaped TiN beams having a width of 6 µm [8], see Figure 2. Due to their small thickness of only 50 nm particular care was taken for the suspension of TiN beams, which was accomplished by critical point drying with supercritical CO_2 after the last etch step in order to avoid beam stiction ('static friction') to the ground plate [12]. Electrical testing was performed after stealth dicing the sensor chips from the full wafer [13] and contacting ∅ 80 µm bond pads with a flexible cable via flip-chip bonding.

2.2. in-vitro Tests

Investigations of biostability were performed *in vitro* by exposing the sensor chips to different model solutions and determining the loss rates by comparing selected geometrical parameters in their initial and final state. Figure 2 displays a microscopic view into the MEMS cavity showing the full X-shaped actuator beam that is situated 2.5 µm above the ground plate. Length *L* and width *B* were determined by optical microscopy with a precision of ca. 1 µm before and after the corrosion test.

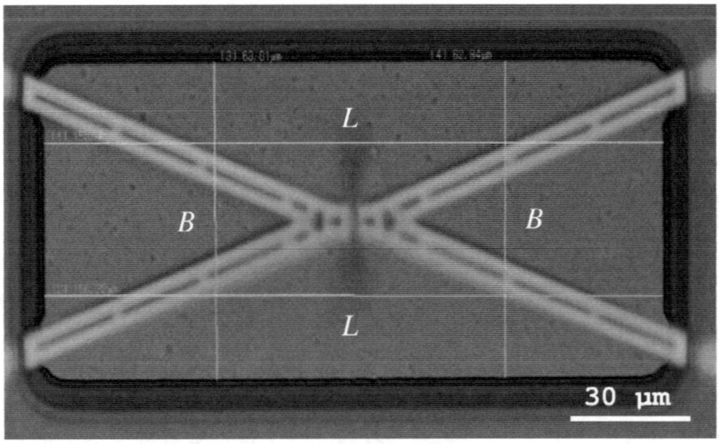

Figure 2. Sensor MEMS cavity showing the X-shaped beam at a height of 2.5 µm above the ground plate that both were prepared from TiN. Green lines indicate the distances *L* and *B* that were determined prior and after corrosion testing.

A possible reduction of top layer thicknesses was examined by spectroscopic ellipsometry. SE measurements were done in the wavelength range of 190-800 nm using an automated, small-spot (14×28 µm^2) spectroscopic ellipsometer with a 71° angle of incidence on the selected area of the chip. The determination of thicknesses (Pas (d_{pas}), SiO_2 (d_{ox}), TiN) of multilayer stack was performed. Surface roughness R_a was determined from topographic AFM scans.

For simulating the body fluid a model solution of isotonous saline PBS-buffered to pH = 7.4 and constantly held at 37°C was applied, into which the sensor chips were completely immersed. Optical and electron microscopy were applied for the inspection of corrosional damage that occurred for some sensor chips during electrical testing in glucose-containing electrolyte.

2.3. in-vivo Test

In order to evaluate the corrosion resistance under realistic operation conditions, sensor chips were also implanted into the abdomen of a volunteer, where they remained for three month before they were explanted and examined. Chips were inserted into polysulfone tubes with a cut-off of 300 kDa and thus allowing for the safe access of corrosion-active components from the body fluid. Corrosion effects were investigated by the methods outlined above by determining variations of thicknesses and other geometry parameters by comparing initial and final state. Disinfection of samples before and after the test was performed by immersing them in ethanol (>70%).

3. Results and discussion

Sensor chips were first subjected to model solutions without bringing them to operation, i.e. no voltage pulses were applied as required for the viscosity measurement. Sensor chips were simply stored in isotonous saline at 37°C and regularly taken out for microscopic inspection. No increase of geometry parameters *L* and *B* of the MEMS cavity could be determined within the measurement precision after six weeks, $\Delta L \approx 0$ und $\Delta B \approx 0$. Similarly, no significant variations of thickness of the passivation layer d_{pas} or surface roughness R_a of the TiN ground plate could be stated.

Comparable tests were performed with functionally operating sensor chips that were subjected to some 100 ms long DC voltage pulses of 3..4 V at regular time intervals between 10..300 s. Also in this case, no variations of L and B parameters above the significance level could be determined. However, a complete failure of the sensor operation was frequently observed during these tests that could in some cases be traced back to a corrosion of planar interconnects. The phenomenon is illustrated in Figure 3 by optical and electron microscopic views of corroded metal lines. A disruption of the metal line can clearly be recognized that caused parts of the electrical circuit to become disconnected from the voltage source.

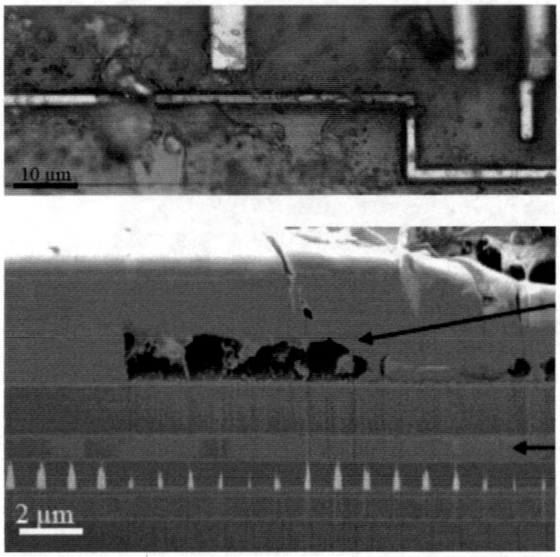

Figure 3. Optical microscope view on chip section showing corroded metal line as yellow line that is greenish coloured in the corroded state (top).

FIB REM micrograph of corroded conduction line of the same area as the top figure showing that the metal layer TM1 below the passivation layer Pas was severely damaged (bottom).

An analysis of the effect revealed that the corrosion originated from a defect in the passivation layer, by which an electrical contact between the electrolyte solution and the metal line TM1 was introduced. It may be suspected that the corrosion process is related to the flow of electrical current in the metal line, since such phenomena were not observed for non-operating sensor chips.

Evidently, such defects have to be completely suppressed for biosensor chips that shall reliably operate in an electrolyte-containing environment. A modification of the top-most layers of the BEoL stack was accordingly tested for solving the problem. Figure 4(a) displays a cross section of the layer architecture from the standard process, where no CMP is applied after depositing ILD4 on TM1. The finally obtained passivation surface then shows protrusions in all those areas, where lateral interconnects made from TM1 can be found below.

It could be assumed that the observed corrosion of metal lines initiated in the vicinity of protrusions being particular prone to mechanical damage upon inadequate handling of sensor chips. Defects would then establish the first contact between electrolyte and TM1 and would act as starting point for the corrosion that would progressively remove increasingly distant parts of interconnect. Figure 4(b) shows the modified BEoL stack, where ILD4 is supplemented by a SiO_2 layer that was subsequently planarized by an additional CMP step. For this architecture, metal interconnects from TM1, firstly, do not cause topographic protrusions on the surface and, secondly, exhibit a larger distance to the surface that stays in direct contact with the bio-milieu. Both effects preserve the metallic interconnects of TM1 effectively from the corroding environment.

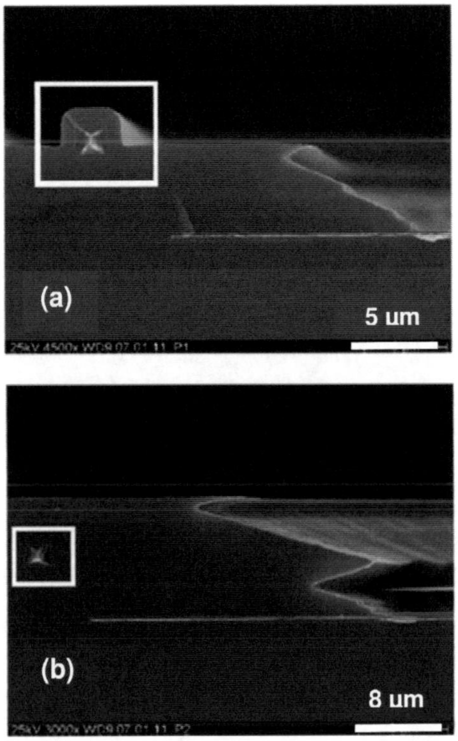

Figure 4. SEM cross sections of two sensor chips close to MEMS cavities showing SiO_2 etch slopes and TM1 interconnects (white boxes). Both chip architectures are finished by a top-most passivation (Pas) layer from SiON. The stack given in (a) has been prepared by the standard process yielding topographic protrusions in the area of TM1 interconnects. In Figure (b), however, the ILD4 layer process was supplemented by depositing another SiO_2 layer and its subsequent planarization by CMP. In this configuration, the Pas layer that stands in direct contact with the bio-milieu is perfectly flat and is less prone to in-creeping electrolyte.

In order to examine the biostability under realistic conditions, *in vivo* tests were performed without supplying a voltage to the sensor chips and without operating them as microviscosimeter. For this purpose, sensor chips were inserted in polysulfone tubes and implanted into the abdomen of a volunteer. Samples were fixed at a depth of 10..20 mm below the skin surface by a medical fathom to prevent migration. Chips stayed for three month in the body and were finally explanted. No significant variations of surface roughness R_a induced by bio-corrosion was observed for the TiN ground plate.

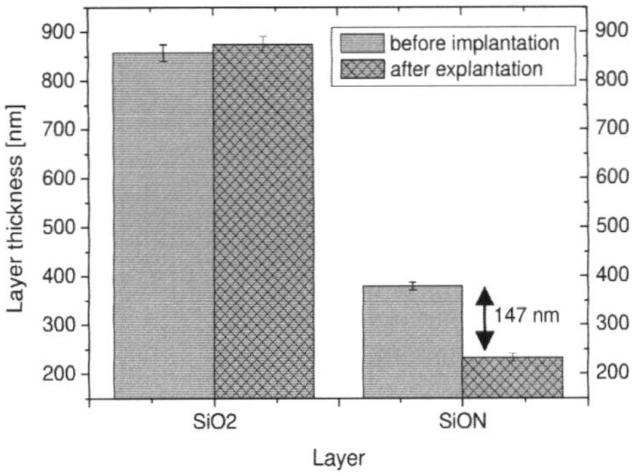

Figure 5. Ellipsometrically determined thickness of top-most SiO_2 and SiON layers within the BEoL stack of *in vivo* tested sensor chips. A loss rate of 147 nm was derived for the SiON Pas layer that stayed in direct contact with body fluid, while the SiO_2 layer below it remained practically unaffected.

Figure 5 gives the thicknesses d_{ox} and d_{pas} of top-most BEoL layers by comparing them before and after implantation. It can be recognized that the SiO_2 layer remained unaffected within the limits of precision of the measurement. This may be understood from the protecting SiON layer situated on top of it. The thickness of the latter, however, has been diminished by 147 ± 8 nm during the experiment. The loss rate of about 50 nm/month compares well to the results of a previous *in-vivo* study that was related to the development of an implantable retina chip [9].

4. Conclusions

The investigations of a biosensor chip intended for long-time implantation presented here revealed no measureable loss rates for exposing the chips *in vitro* for six weeks to model solutions of the body fluid. An improved back-end architecture has been developed to avoid the corrosion of metal interconnects that was frequently observed for electrically driven sensor chips. As a first estimate for a possible life time, it may be deduced from an *in vivo* test that sensor chips may operate for a time span of months in the human body.

5. References

[1] Hierlemann A 2009 *Procedia Chemistry* **1** 5-8
[2] Graham A H D, Robbins J, Bowen C R and Taylor J 2011 *Sensors* **11** 4943-71
[3] Ehwald R, Ballerstadt R and Dautzenberg H 1996 *Analyt. Biochem.* **234** 1-8
[4] Birkholz M, Ehwald K-E, Ehwald R, Kaynak M, Borngräber J, Drews J, Haak U, Klatt J, Matthus E, Schoof G, Schulz K, Tillack B, Winkler W and Wolansky D 2009. In: *Mikrosystemtechnik Kongress 2009,* ed H Seidel, *et al.* (Berlin: VDE Verlag) pp 124-5
[5] Schultz J S, Mansouri S and Goldstein I J 1982 *Diabet. Care* **5** 245-53
[6] Ballerstädt R and Ehwald R 1994 *Biosens. Bioelectr.* **9** 557-67
[7] Birkholz M, Ehwald K-E, Fröhlich M, Kulse P, Basmer T, Ehwald R, Guschauski T, Stoll U, Siegel H, Schmaderer S, Szeponik J and Zahn D 2012. In: *Sensoren und Messsysteme 2012,* (Nürnberg: GMA ITG VDI/VDE) pp 177-87
[8] Birkholz M, Ehwald K-E, Kulse P, Drews J, Fröhlich M, Haak U, Kaynak M, Matthus E, Schulz K and Wolansky D 2011 *Adv. Func. Mat.* **21** 1652-6
[9] Hämmerle H, Kobuch K, Kohler K, Nisch W, Sachs H and Stelzle M 2002 *Biomat.* **23** 797-804
[10] Birkholz M, Ehwald K-E, Wolansky D, Costina I, Baristiran-Kaynak C, Fröhlich M, Beyer H, Kapp A and Lisdat F 2010 *Surf. Coat. Technol.* **204** 2055-9
[11] Knoll D, Ehwald K-E, Heinemann B, Fox A, Blum K, Rücker H, Fürnhammer F, Senapati B, Barth R, Haak U, Höppner W, Drews J, Kurps R, Marschmeyer S, Richter H H, Grabolla T, Kuck B, Fursenko O, Schley P, Scholz R, Tillack B, Yamamoto Y, Köpke K, Wulf H E, Wolansky D and Winkler W 2002. In: *IEDM Techn. Digest,* pp 783-6
[12] Kulse P, Birkholz M, Ehwald K-E, Bauer J, Drews J, Haak U, Höppner W, Katzer J, Schulz K and Wolansky D 2012 *Microelectr. Eng.* in press
[13] Birkholz M, Ehwald K-E, Kaynak M, Semperowitsch T, Holz B and Nordhoff S 2010 *J. Optoelectr. Adv. Mat.* **12** 479-83

Acknowledgments

The investigations presented here were performed within the development project of an implantable glucose sensor funded by the BMBF under contract number 16SV3934. We thank our co-workers from IHP pilot line for the preparation of sensor chips, as well as our co-workers from process research, in particular, U. Haak and K. Schulz for photolithography, D. Wolansky and A. Scheit for metal deposition and J. Drews for MEMS etching and critical point drying. Many thanks is also to G. Fritsch, Frankfurt (Oder), for implanting the samples, R. Ehwald, Berlin, for polysulfone tubes as well as Invent Diagnostica GmbH and In-vitro DiagnostikNet, Berlin, for their support of this work.

Institute of Physics
Dirac House, Temple Back
Bristol BS1 6BE UK

CURRAN ASSOCIATES INC
proceedings.com